Rowland Jones

The circles of Gomer

An essay towards an investigation and introduction of the English as an universal

language

Rowland Jones

The circles of Gomer
An essay towards an investigation and introduction of the English as an universal language

ISBN/EAN: 9783337281397

Printed in Europe, USA, Canada, Australia, Japan

Cover: Foto ©Andreas Hilbeck / pixelio.de

More available books at **www.hansebooks.com**

CIRCLES of GOMER,

OR, AN

ESSAY

TOWARDS AN

Inveſtigation and Introduction of the E N G L I S H, as an UNIVERSAL LANGUAGE, upon the firſt Principles of Speech, according to its Hieroglyfic Signs, Argrafic, Archetypes, and ſuperior Pretenſions to Originality; a retrieval of original Knowledge; and a re-union of Nations and Opinions on the like Principles, as well as the Evi dence of ancient Writers;

WITH AN

ENGLISH GRAMMAR,

Some Illuſtrations of the Subjects of the Author's late Eſſays, and other intereſting Diſcoveries.

By R O W. J O N E S, Esℚ.

THE

CIRCLES of GOMER.

ABALLABA or Apelby, and Abala, a vale and town on the river Eden, W-mor. another of the Troglodites by the Red Sea, and the name of feveral other places in Ire. and Wales, fignify the water place or vale part; and in a fecondary fenfe, as rivers, feas, and waters, were the ufual limits and extents of provinces and diftricts, the loweft and fartheft place; alfo Bala, inflects into fale or vale, and afal, apple, or the vale produce, fo named after the place of its growth, like the wines of Oporto, Champaign, Burgundy, Mountain, Rhenifh, and others. The name Belatucadro, infcribed on coins found here, fignifies the keeper or warden of the fartheft fide or confines water; as does Kirkby Thor, fituated on the fame river, mean the confines or water-gate part; and the infcription Thur gut Luetis, on another fmall filver coin there found, and fuppofed to be the Saxon deity Thor, to whom our Thurfday was dedicated, Thor gate let, or the water-gate let, or its keeper; it being probably a payment of toll or duty to him for the liberty of pafing and repaffing his gate on fair and market-days; and to whom, on the expulfion of the Druids and Britains into the northern countries, the fuperftitious Roman, or the ftill more Pagan Gothi-Dane, might fet up the idol: it having been an ufual practice among the Hebrews, as the Greeks to deify their noble anceftors; for the Gothic nations fcem to have not have ftrictly adhered to the worfhip of the true God, like the true patriarchs; and though no fuch Gothic infcription has been difcovered, the Romans have ————, may————, with a view to traducing the Druidical religion, and which of the proper notes their great attachment to it. Abana, a river of Cœlo-Syria, the increafing

B　　　　　　　　　　　　　spring

ſpring part. Abantes, or Trinobantes, of Ionia, Greece, and Britain, the incloſing ſpring water ſides, parts, bottoms, vales, or dwellings. Abaris, or Abas, a Scythian poet, who dwelt in Greece, the water part ſide, a bard, or the regiſter or recorder of the Druids, Abbots, or Barons in Britain, at their cells, leets, or palaces on the boundary water part, like Abdias the ſteward of Ahab and Abdon of the Judges of Iſrael. Abdera of Spain, the ſpring water part. Abeda or Bede N-hum. the ſpring water part ſtation or bed. Abel, a part from, or ſon of the water place or light. Aber, the name of many places in England, Scotland, and Ireland, at the mouths, confluence, or diſemboguement of rivers ; in its primary ſenſe ſignifies the ſpring water part ; ſecondly, as the water places and harbors of ſeamen, an haven or harbor ; and thirdly, from the nature of the ſcituation, a confluence of water. Aberbroth, Scot. the ferry or port water part or harbor. Aber Conwy, N-Wales, the confines or incloſeing ſpring water part or harbour. Aberdeen, Scot. the harbor on the diviſion ſpring water. Aberduglethi, or Milford haven, S. Wales, the haven of the two ſide ſpring water places, or of the two ſwords, as both cletheu and ſword in their primary ſenſe ſignify the ſpring water part, whoſe hieroglyphic is the male part of propagation or generation. Aberford, or Aberforth, near the Roman road and river Cock, Yorkſ. the water part way or ford. Aberfraw, N-Wales, the ſpring water part harbor, or haven. Abergaveny, or Gobanium on the Wye and Goveny, S-Wales, on the confines or incloſeing ſpring water parts. Aberguain, S-Wales, the incloſeing ſpring water part or harbor. Aber Meneu, N-Wales, the edge or ſurrounding water part, and as the edge of any thing is narrow, the narrow water part. Abernethy, near the conflux of the Ern and Tay, Scot. the ſide incloſing water part. Abington, or Abandon on the Ouſe, Berks. the incloſeing or confines ſpring water part ; and, as abbeys, cells, convents, monaſteries, priories, and churches derive their names from their being ſcituated on the ſea or river ſides, the abbey town. Abinger, Surry, the incloſing ſpring water part, border, or head. Aborigenes, or Pelaſgians of Italy and other places, who, according to Virgil and others, dwelt on the tops of mountains and other high places before the draining of the flat countries by rivers and other drains by Hercules, ſignify the Phrygian or the fartheſt or lowermoſt water confines or borders. Abotſbury, Dorſet, in its various meanings, like the city Abotis in Egypt, ſignifies the ſurrounding water ſide, ſpring part, habitation, burrough, or an abbot's barrow or reſidence ; and from the decreaſe of ſprings, increaſing

increasing the sea, and their junction, came abate, abet, and abut. Abone, Aboyn, Aboy, Aubon, Aufon, or Avon, appellative or common names of rivers in Som. Wilts. N-ham. Dorfet, Suff. Norf. War. Gloc. and moſt other parts of England and Ireland, fignify the furrounding or inclofeing, or fide fpring water part or fountain. Abravanus or Abervan, Scot. the fpring water part, port, or haven. Abrevicum or Berwick on the Tweed, the fpring water part, harbor, ſtreet, or fortification. Abretana, and Abretanus in Myfia, and one of Jupiter's names, at the water part fide or coaſt. Abus, Humber or Chumber, Yorkf. the fpring water lower or fea-fide part, or an æſtuarie; it being fo called from Abia the daughter of Hercules, or a Cumbrian colony, which opened and fettled on that river. To thefe names of places and perfons, which are the primary and moſt perfect models of all common names of things, the Englifh dialect owes its origin; as in the following names, viz. Abaft, the fpring part off the fide; Abafe and Abaſh, the fpring part low, which the reverfing the letter b hieroglyfically expreffes; Abate, the fpring part at the fide or fea; Abba, the fpring part, as Mama is the furrounding water part; Abbey, the fpring water part or abode; Abbreviate, the fpring water part at the fide water; Abdicate, from the fide or confines keeping; Abed, Abet, and Abide, at the fpring part fide; Aberring, from the inclofeing water part; Abhor, the fpring part behind the border or from the border; Able, the fpring water place; Aboard, on the fide furrounding or inclofeing water part or fea veffel; Abode, the furrounding water part fide; Abolifh, the furrounding fpring water place low or leffened; Abominate, at the furrounding water part edge; Above, the furrounding water part fpring or fcurce; Abound, the furrounding fpring on the fides; About, the furrounding fpring water fide; Abreaſt, the water part lower fides; Abridge, the water part edge; Abroach, at the furrounding water part action; Abroad, from the road, neighbourhood, country, or furrounding water part; Abrogate, from being demanded; Abrupt, the fpring water part up to the furface; Abcefs, Abfcind, Abfcond, and Abfence, the part below the lower or fide confines; Abfolve and Abfolute, the part below or at the fide of the furrounding fpring water place; Abforb, the part below the orb or border part; Abſtain, the part below the inclofeing coaſt; Abſtract, the part below the water confines fide; Abfurd, the part below the fpring fide; Abufe, from ufe, or from the fpring part fide; Abut, the fpring part fides, which abut; Abyfs, the lower part.

Ac, Aue, or Aqua Solis, Bath, Borth, Baden, or Ackman-
chefter, Som. the fide or hot fpring water part, and the confines
water part city; and from fick people's attending there. the
Acke, or Sick Man's City; Borth and Aqua Solis being both
expreffive of the furrounding light part as well as the furrounding water part. Aca, Acabe, Acalene, or Acabis, in Mefopo-
tamia, Macedonia, Egypt, Lybia, and Spain, the water confines, the water confines part, end, or fide, or the lower
water fide. Achaia in Greece, the inclofed, or water inclofed
par, or fields. Achelandia or Aucland, Dur. the fpring water
fide bank, land, or poffeffions. Acheron or Acheron, a river
of Greece, the inclofing fpring water or river. Acmedie or
Hemedie iflands near the Orcades in the German ocean, the
furrounding water or fea parts or iflands. Acquitania and Lacce-
tania, a part of Gaul and Spain, on the confines water fide, or
coaft, or below the confines or inclofing water. Actium in
Epirus, the confines water fide. Acton, and Acton Burnel,
Glec. and Salop, the water or fpring water place fide town.
Ackman Street, the Wicci fpring water confines, Walking
and Aking fireet. Hence Accede and Accefs, the confines water fide, or banks, Accelerate, the running water
place; Accept, at the confines water part; Accident, below
or under the confines water fide; Accompany, on the fame
furrounding water part; Accompt, its confines; Accord, at
the water border; Accefs, the water confines lower fide, or
edge; Accrue, the fpring water confines, or a place of growth;
Accufe, below the water border; Accule, below the confines fide, dominion, its limits and dominion; Ace, the
fpring water fide; Ache; Acre, the inclofing or dividing
border, the water inclofing part, which might have been
a perches broad and 40 in length, with fufficient depth, al-
lowed on the firft divifion of countries, a chief for, making,
imparting, and defending fo much the bounds, Acu and
Actum, the confines water fide, Acuate and Acute, the confine fpring fide or edge.

Adam or Adham, the divided or created part, man, houfe
or houfes. Ad adam or Ad amram, near Camulodunum or
Malden in Effex, on the fide inclofing fpring water part.
Adre Iw. the fide or divifion water part. Adeling or
Athelney, Som. the fide inclofing water place, vale, or holding. Adeling, a title of nobility, and of the eldeft fon or heir
apparent of England, fignifies a holder of the Lords, ifland
or England, from whence nobility and fovereignty feem to be
derived. Aderbury, Wilts, on the fpring water part fide.
Ades, Hades, or Dis, the lower part, hell, or its prince, who

Is suppofed to be Pluto king of Spain. Adgebrin or Yverin, N-hum, on the confines water part or fpring edge. Adington or Haddington, Wilts, N-ham, Surry, and Scot. the inclofing water fide town. Adfield or Hatfield, Herts. and Yorkf. the fide dwelling-place, field, or village, which perhaps was at firft the acre. Adminius, the fon of Cunobelin, and chieftain of the Trinobantes, at the furrounding water fide or edge. Adriaticum or Adrianum, a fea betwixt Italy and Dalmatia, the fide furrounding water part or fea. Adres or Bardfey Iflands, N-Wales, the fide furrounding water parts. Adurni portus or Ederington, Suff. the fide inclofing or confines water part or town. Adwick in the Street. the fide fpring confines, ftreet, fortification, or village, which Adwick fully expreffes, without the addition of in the ftreet. Hence, as a the earth is divided from e the water by d the fide, Add, Addition, and Adequate fignify the fide or equal divifion; Addle, the fide water place, which is at the water edge, and under it is barren; Adeling, the inclofing water place fide, holder, or nobleman; Adrefs, below the fide water part; Adept, at the fide water part; Adhere, the fide water part, or at the fides; Adjoin, the fide inclofing circle; Admiral, on the fide furrounding water or fea; Admire, the fide furrounding lights; Admit, at or to the furrounding fide or border; Admonifh, at the lower furrounding water part or fea fide cell or monaftery, where people were ufually admonifhed; Ado, the fide circle of either fort; Adolefcence, the circle fide below the confines water place; Adore, the border fide, which were the places of adoration; Adorn, including the furrounding fide; Adrift, flowing at the fide water part or fea; Adry, at the fea water; Advance, the fide lower part; Advantage, the fide part taking; Advent, the fide fpring into; Adverb, the fide or added fpring part; Adverfe, below the fide fpring water; Advert, at the fide fpring of generation, munition, fpeech, hearing, feeing, &c. Advife, at the lower or leffer fpring; Aduft, at the fpring part on the fide; Adulterer, at the fpring part on the fide man or married perfon; Adumbrate, the fide furrounding part; Advocate, at the fide circle or calling fpring part; Adverfion, the calling fide or divifion; Adrit, the fide fpring at the loweft.

Æas and Æacus, a water part or ifland, and an ifland chieftain. Æbuda, Æboda, or the Ebudes, the water furrounding parts or habitations. Ædui, Ædri, or Burgundi or Gallia Narbonenfis, who probably were the in migration from the Mediterranean fea on the banks of the Rhine, the upper fpring parts, dwellings, or dwellers; or any fon

of Titan and Terra, whom Gods Briareus, and men Ægeon, named according to Homer, fignifies on the water fide or edge, or the Weftern or inclofing water confines. Ægypt, the edge, or confines water part or firft inhabited country. Ælward and Ælfred, or Ælfrith, the water place fide, ford, or warden. Æthiopia, fouthward of Ægypt in Africa, the edge or fartheft water part. Ætona, Berks, on the water fide town, or the inclofeing border water part. Hence Æra, the upon or furrounding part; Aerial and Ætherial, on the fide or furrounding fpring or water. Ætna, for Ætan, the fire.

AFFRANIA, a fhamelefs woman, the inclofeing water parts. Affrica, the water part, or hot confines or country. Affrim, the water part run, or border. Affrodifius, leffening the fide, furrounding or inclofeing water. Hence Afar, the water part; Affable, the water place part; Affair, the earth part, or on the part; Affect, the water part at; Affiance, the the part on the fide; Affirm, on the furrounding part; Affix, the part to the confines water; Afflict, the part from flowing; Afford, the way, ford, or furrounding water part fide; Affray, on the water part action; Affright, the part on the confines water fide; Affront, the front, or the water part on the fide; Affay and Affy, the parts fpringing, growing, or acting together; Afloat, the furrounding water place part or at the flood; afoot, the paw or part at the furface; Afore, on the furrounding water; After, off the water fide or part.

AGAG, the inclofeing or confines water chief. Agar, the inclofeing water. Agathyrfi, or Picti Agathyrii on the borders of Scythia, the confines or inclofeing water lower fide or upper part. Agelocum or Littleburgh, Linc. the furrounding water place, confines, or edge, or a burrow on the fide water place. Agefilaus or Pluto, on the lower confines place. Aglaus, the fpring water place edge, or confines. Agrippa, Italy, the border water part. Agyllcus, the fon of Hercules, who drained the midland countries and founded the Gauiifh nations, the fpring water place confines. Hence Again, on the inclofeing water edge; Againft, on the inclofeing water lower fide; Age, the inclofeing water action, oge, or circle; Agent, on the inclofeing water fide; Agitate, the confines water fide part or its action; Ago, the circle or action from; Agrarian, inclofeing or dividing the countries or boundaries; Agree, inclofeing the water confines; Ague, the fpring water confines. Ah and eh, the a and e found as interjections.

AIBRIDGE, Effex, the fpring or flowing water part edge, or bridge. Aich or Aix, Aquitan and Germany, the fpring or

8

or flowing water confines. Aidon or Haidon, Yorkf. and N-hum, the fpring, or high fpring water fide or town. Ailfbury, Ford, and Ham, Bucks, Kent and Norf. the fpring or flowing water place fide burrow, ford, vale, or dwelling-place. Aire, Scot. the fpring or flowing water. Hence Aid, the fide flood or fpring; Ail, from flowing; Aim, the furrounding flood or water edge; Air, the flowing fpring or fire; Aifle, the ifle or fide flood or water place; Ait or Eyght, the fide flowing fpring water part.

ALEMANNI of Germany, the fpring water place fides, alleys, Gaulifh, or vale parts; and, in a fecondary fenfe, the hilly parts or banks which inclofe water places, like the hills and dales of Scotland, which are thence named Almond. Thofe of Germany are by hiftorians fuppofed to be a motley race of Sueves, Gauls, and others, who, in the beginning of the third century, feated themfelves betwixt the Danube, Upper Rhine, and Meyne. Alani, Glani or Clani, Alaun, Alen, Aln, Alone, and Alway, Corn. Dorfet, Hants, War. N-ham, Lanc. Cumb. N-hum. Scot. Brittany, and the banks of the Danube, the inclofeing fpring water place, banks, or vales, or the Britifh Glyns, Glens, or Clans; thofe of Germany being probably a branch of the Cymmeri, Cumberi, Umbri or Chumberi, and Cumata or Comata, who inhabited almoft all the vales and inland parts of Europe. Alaw, N-Wales, the fpring water place or river. Alaric, the upper country chief, or king. Albany, Alben, Albion, or Scotland, the upper weftern water place or ifland, as is Britain the lower water part, and the high or hilly ends or banks in a fecondary fenfe. Alba-land or Alba Ecclefia, N-hum. the fpring water place bank fide, church, or poffeffions; the Englifh land being compofed of lan, a water bank or church-yard, and t or d, the fides and poffeffions. St. Alban's, Wer-linga or Watlinga ceafter, or Verulam, or Jan, Herts, the fpring water place confines or banks. Albenich, Scot. the the upper hill or high land. Albemare, the high or upper mere part. Albrighton, Salop, on the confines water part town. Albury and Alchefter, Surry, Oxon, and War. on the fpring water place, dwelling, burrow, or city. Alcmena, the mother of Hercules by Jupiter, the furrounding water place or fea confines, ports, or havens. Alcuin, on the confines fpring water place. Aldburrow, Suff. the water place fide burrow. Aldenham and Aldenbargh, Herts and Weftphalia, a burrow or vale on the water place fide. Alderley, Gloc. a place or palace on the water part. Alderman, a man on the water part. Aldermafton, Berks, a town on the middle or furrounding

ing water part. Alderminster Were. a station or monastery on the side water part. Alderney or Riduna Island, a part in the high water or sea. Aldingham, Lanc. a vale on the confines water side. Aldport, Lanc. the side water place, port, or the old port. Aldwin, on the side spring water place. Alvetum or Dundee, Scot. on the side or division spring water place. Alerton, Yorks. the spring water place town. Alford, Chef. the spring water place, ford, or way, or the highway. Alfred, on the water place side, or Britain. Alfreton, Der. on the water part town. Alice, the side water place. Allington, Kent, on the incloseing water place or bank town. Allendorf, on the Upper Rhine, the water bank, or border gate part. Allobriges, of the Viennois, Savoy, and other parts of Narbon Gaul, whose capital was Vienna on the Rhosne, and who seem to be of the same origin as the Venta Belgarum of Gaul and Britain, signifies the surrounding spring water place, vale, alley, Gaulish upper water parts, or Phrygian confines. Alnwick, N-hum. the incloseing spring water place, street, or fortification. Aloa, Scot. on the surrounding water place or sea. Alon, Cumb. the incloseing water bank or lane. Alpes of Italy, above the low parts, or the hill parts. Alresford, Hants. a ford or way on the side water place. Alsten Moor on the Tine, Cumb. a moor below the confines spring side. Alt and Altena, Lanc. and on the Elbe, the spring water place side, bank, or hill. Alterynis, Heref. on the incloseing water part or island. Alvingham, Linc. on the confines water place vale, or dwelling-place. Alwen, N-Wales, the high incloseing spring water place. Hence Alack, from being high; Alacrity, on the confines water place, or on the wing; Alarm, on the surrounding parts; Albeit, on the side part; Alchymy, the high art; Alcoran, on the high borders; Alcove, a covering upon; Alder, on the water part; Alderman, on the water part man; Ale, the spring water; Alert, at the spring water place; Algebra, on the confines water parts; Algid, on the confines water place side; Alien, an ally one or in another alley, street, or country; Alike, the water place likes, or shadow; Alive, on the flowing spring; All or Ol. the surrounding or circle water place; Alb, on the high water; Alledge, on the water place edge; as alledging on the water edge before a purgation by water ordeal; Allegiance, a tye within the country or water confines; Aldrich, on the high flood or ditch; Alley, on the water place; Alliance, the being on the same confines, water ally; Allodial, a holding a member of the ancient property, from the Welsh Alod, a member; Allot,

on the furrounding water place fide; Allow and Law, on or
the cuftom of the alley; Ally, on the fame alley; Almanac,
on the furrounding water place or moon action; Almoft, the
furrounding water place or fea loweft fide; Alms, the furrounding water place fide or palace; Alocs, the high age or
lower or paft circle; Alone, all one; Along, the inclofing
water place; Aloof, the furrounding water place part; Alpha
and Alphabet, on parts and beafts; Already, on the water
part fide; Always, on the fpring; Aliter or Alter, the alley
or Gaulifh poffeffions, which were different from the Romans,
and Alitera or Ceres, the goddefs or chief of the corn.

AMANUS, Amonus of Afia, Spain and other parts, the furrounding parts, water parts or mountains. Ambarri or Am-
brones of Gaul, on the furrounding water parts or fea fide. Am-
berley, Suff. the furrounding water part place or the amber place.
Ambiorix, the furrounding water part chieftan or king. Ambog-
lana or Amblefide, W-mor. the furrounding water part fides
or banks. Ambrefbury Wilts, the furrounding water part
fide burrow. Ambry, Heref. the furrounding or boundary
water part. Amerfham or Agmondefham Bucks, the furrounding water fide or edge vale. Amifia or Embden Germany,
the lower furrounding water part fide or vale. Amond or Au-
mund Scot. the hill furrounding water or fpring fide. Am-
phitrite the wife of Neptune, the furrounding water part or
fea. Ampthill Bed. the furrounding water part fide hill.
Hence Amain, the main fea; Amafs, the furrounding or inclofing fides parts, or countries; Amber, from its being found
on the fea fide; Ambient, encompaffing or furrounding the
parts on the fides; Ambit, circuit or the fide furrounding part;
Ambition, on the fide furrounding part; Ambulation, on the
furrounding fpring water place fide; Ambufh, the furrounding
fpring part fide; Amen, for or within the bounds or heaven;
Amenable, within the furrounding water place; Amend, for
or about the end or infide; Amerce, the furrounding water
fide; Amifs, below or at the fide of the furrounding water
part; Amity, for the fide or the furrounding water fide;
Among, within the furrounding water circle; Amongft, within the furrounding water circle lower fide; Amoer, for man
or furrounding water circle; Amount, the things in the furrounding circle; Amphibious, the two fides or furrounding
parts or boundarys, which belong to neither diftrict; Ample,
the furrounding water place; Amputate, the furrounding
fpring part at the fide; Amufe, the furrounding fpring fid..

ANALE or Longford Ire the incloſing fpring water place or
the water bank way or ford. Anas and Anabis Spain, the

C   lower

lower inclofeing fpring part. Anacreon, the inclofeing fpring water one. Ancalites or Henly Oxon, the inclofeing fpring or ancient water or dwelling place. Ancafter or Crococalana Linc the inclofeing fpring city or the border water bank. Andates the goddefs of victory, the inclofcing fpring water part fide. Anderida or Newenden Kent, the part on the water part or the vale or foreft on the inclofeing fpring. Andernefs Lanc. a promontory on the water part. Andover Hants, on the boundary river. Andradfwald Suff. on the water part fides wild or wood. Andraftes, on the water part lower fide. Aneftie, Auuftie, Anfty or Hanfty Herts and Yorkf. the inclofeing fpring fide or poffeffion or the ancient poffeffions belonging to the river warden for the fupport and defence of the boundary. Anglefey or Mon. N-Wales, the Angles ifland or on the furrounding weftern water ; it being an ifland of the Eaft Angles or Wenta Ifcenorum, who made their weftward migrations thither, thro' the inland countries of the Angli Mediteranei, the midland Angles or the country of the Cornavii. Angloen Germany, the inclofeing water place corner. Hamble or Homlea the furrounding fpring water or home place. Homlea fays Bede runs thro' the country of the Jutes, the Kentifh men, and the Ifle of Wight, and falls into the Solente, the fea fide water place or bank on the Hampfhire coaft; and the Angles country is that called Angulus fcituated between the Jutes or Kentifh men and the Saxons or Suffex Surry and Effexmen. Then fays Camden if the right meaning of Engelbert and Engelbard were difcovered it would help us to that of England and Angli. The meaning of thefe names feems to be Engli Britain, or Englifh water parts or poffeffions, as is England, Englifh land and Angli, Engli or rather its firft denomination Ingli, a place inclofed or confined by water or an ifland, different from Angles or corner ifland or the Ifceni confines water place, as being furrounded only by a part of the Thames on the Effex fide, and the Angles fea coaft according to the original meaning of the radical in the Welfh as well as the Englifh dialect of the Celtic, and what other people or country can be underftood from this defcription muft be difficult to find out ; for as to Angulus, there being then neither people or country of that denomination, befides thofe of England it feems to be a Latin name of Bede's own coining agreeable to his general plan of introducing a new fort of grammatical language different from all the vulgar dialects of Britain, and perhaps of any other country ; but was the text fet right it feems probable that this well meaning leading hiftorian would appear to mean no other than the Angles of Britain. Especially fince

Bede

Bede and other hiftorians confider the Angles and Saxons as diftinct nations. Angre Chipping Eff. the corner or angles confines
water parts. Angus Scot. the inclofeing water confines   Antia
Italy, the fide fpring.   Ankam or Ad Anfam and Anker Lil.
Norf. War. the furrounding or inclofeing fpring water confines
or border. Anlaby Yorkf. the inclofeing fpring water place or ancient part or habitation. Annan and Annandale Scot. the inclofeing fpring or ancient vale or dwelling place.   Anticaria Spain,
on the fide borders. Antivæfteum, Penwith, Bolerium or Lands
end Corn. on the lower water or weftern fide coaft, the wa-
•·r fide end or on the furrounding water place part, a Roman
name compofed of anti, and the Englifh name weft or we-ef-tu,
the water lower fide in the dialect of the Veneti or Wenta,
who were fome of the firft poffeffors of Britain, and from whom
the prefent Englifh language feems moftly to derive its origin,
as appears on examination of the conjugating verbs, particles of
conftruction, the names of perfons and places, and manner of
compounding both.   Antona or Northampton, the inclofeing
fpring town.   Antport or Hanton port Hants, the inclofeing
fpring fide port.   Antrim Ire. the inclofeing fpring fide border or rim.   Antwerp, the inclofeing fpring fide water part.
Anubis, an Ægyptian god of the fhape of a dog, the inclofeing fpring water part   Hence the Greek Ana, from the inclofeing fpring or without; Anarchy, from the inclofeing
fpring water confines or fides; Analogy, the inclofeing fpring
water place fides, which are alike; Analyfis, from the fpring
water place fides, which are proportionally feparated; Anathe-
ma, from or without the furrounding fide or border; Anatomy,
feparating the furrounding border or body; Anchor, a fhut on
the border water, the letter T or a river running into the fea;
Anchoret, on the inclofeing water fide fhut; Ancient, on the
inclofeing fpring upper fide; And, the inclofeing fpring divifion; Anew, in the fpring; Angle, a corner or confines
water place, as a part of ingle or ongle, a furrounded water
place; Animal, the furrounding fpring place one, Anker,
the water inclofeing or fhut one; Ankle, the leg fhut or joint;
Annals, on the inclofeing fpring; Annex, the inclofeing
fpring water confines; Annihilate, without an illation of
fpring; Annoy, on the fpring circle; Anoint, the fpring in
the furrounding or fide circle; Anno Dom. the furrounding
circle of our Lord; Anon, on the fpring; Another, the inclofeing fpring fide water; Ant, the inclofeing fpring fide, or
ante before, as takeing time by the forelock; Antiquary, an
ancientry man; Antitype, the previoufly divided part; Anti-
pathy, on the inclofeing fpring part fide, which is oppofite to

the

the other fide part; Anvil, the inclofeing fpring on the water place or fea from the knocking and dafhing thereon; Anxiety, inclofed, confined, or ftraitened on the fides; Any, the inclofeing fpring or within it.

APELBY W mor. the water place or fartheft part or habitation. Apenines Italy, the high inclofeing ends or parts. Apewood or Abbots wood Salop, the fpring water part fide, wood or dwelling place. Apelderham Surry, a vale or home on the fartheft water part. Apiæcum or pap caftle Cumb. the furrounding water part caftle. Apiedor, Apletree or Apuldre Kent, the water place vale or fartheft fide, gate or mouth, and appletree as growing therein. Appleton or Nun-Appleton Yorkf. a water place or inclofeing fpring water place town. Apolonia, Ceres or Babelonia, Syria, Chaldea, Afia Minor, Greece, Macedonia, and Italy, the inclofeing water place fides, banks, lanes or lones and lands or lane fides; the inland men having made their migrations from Syria weftward in ftreets on the river banks. Apfcley Gife Salop, the lower fpring water place confines fide. Apthorp N-ham. the fide or confines part water gate. Hence Ap being here a part of or from the earth; Apace is the water part as moving and feparated from the earth; Apart, the fide water part; Apathy, from the fide or fide part; Ape, a part or fon; Apoftate, from the ftate or ftation; Appal, from the water place; Apparel, the part upon; Apparition, on the water part fide or a fhadow; Appeal, from the part higher; Appertain, on the water part or fea coaft; Appetite, at the fide part; Apple, the water place or vale part; Apply, the apple; Appoint and Appofite, on the furrounding water fide part or put; Apprize, from the lower or fide water part; Approach, the furrounding water part meeting together; April, the light flowing water part; Apron, on the water part; Apt, the fide part, as fit or pit is the infide part, which is fitting.

AR and Ararat, the earth, country or upon and on the earth fide. Arabia, the earth dwelling place or the high country part, which, as the bottoms remained covered with water or boggy and undrained for many centuries after the deluge, became the firft inhabited parts, as appears by the ridges, flatts and inclofures on the tops of mountains and hills, and on the mountain of Ararat Noah's ark refted, when the plains of Babylon and Ararat or Armenia were probably under water. Araris a river of Gaul, the lower, flower or fide country water. Araxes a river of Armenia, the upper confines fide water. Arbeiæ or Jerby Cumb. the water part or habitation. Arbon in Suabia, on the upper inclofeing water part. Area Eff. the fhut or confines

fines water. Archeus, the first spring or the first principle or anima mundi of the chymists, which seems to have some correspondence with the springs of language. Archipelago, the chief inclosing water place. Archon, on the water part country. Ardert Ardes and Ardeth Ire. on the side water part. Arden or Dean Gloc. War. on the inclosing water part, vale, forest or wood. Ardmanoch Scot. on the surrounding water or German sea side part or country. Ardmor Ire. on the surrounding water or sea side. Ardoch Scot. on the border water side. Ardudwi N-Wales and Scot. on the division spring water part side. Ardulph, on the spring water place side. Arduthie Scot. on the spring part side. Are or Arus Yorkf. the spring water or river. Areopagus, on the water part confines, sides, streets, towns or cantons. Arecomici of Gaul, the river upper confines or commots. Argentacoxus, on the Kentish or inclosing water confines or coasts chieftan. Arguile or Arguithel Scot. on the inclosing spring water place confines or sides. Argila Ire. on the side water. Argo, on the sea confines. Aricia or Arisia a town of Italy and the wife of Hippolitus, the lower country or the female. Ariconium, Archenfield or Kenchester Heref. on the boundary or inclosing water confines, dwelling place or city. Arith, Erith or Meneg Corn. Camb. Kent, on the water side or surrounding water edge, and in a secondary sense, a free passage thro' a water or a ford. Arklow Ire. on the inclosing water place. Arkley Yorkf. on the inclosing water side. Arlech or Harlech N-Wales, on the water place confines or rocks. Arlington, Harlington or Argenton Mid. Eff. Herts, on the confines or inclosing water town. Armagh Ire. on the surrounding water confines. Armanthwayte Cumb. on the surrounding water or sea side spring water part, haven or harbor. Arme an island in the British sea, on the surrounding water, as an arm on the body or surrounding part or of the sea. Armenia, the surrounding or high country or rocky parts. Armorice or Armorifa of Gaul, on the lower sea, the German being deemed the upper. Arnus Italy, the inclosing spring water. Aretria or Eretria Greece, the side or surrounding water part. Arphaxad or Chaldea, on the division or inclosing water upper part. Arpina Greece, on the inclosing part. Arran or Glota Scot. Ire. on the inclosing water part. Arras Gaul, the lower or lower country. Arrol Scot. on the surrounding spring water place. Arrow War. the surrounding spring water. Arisia Italy, the side or lower water. Artabri Spain, on the water part coast. Arthur, on the side spring water or river Severn. Arverni or Arwerni of Gaul,

on

on the inclofeing fprings or rivers. Arveragus or Arivogus Gloc. the inclofeing fpring water or river Severn confines or chieftan. Arun and Aruns a river of Suff and a foothfayer of Italy, the inclofeing fpring or river fide. Arundele, the Arun vale. Arvonia the infleation of Armonia or Segont N-Wales, on the furrounding weftern water fide or the lower inclofeing water fide. Arufpex Augur or Aufpex a Romifh prieft, on the river fide part, the ufual places of meditation. Arwenack Corn. on the inclofeing fpring water confines or haven. Arwerton Suff. on the fpring water or river town. Arwyftil N-Wales, on the lower fpring fide water place. Hence Arable, the earth dwelling place or poffeffed place; Arbour, on the furrounding fpring water part; Arc or Arch, a covering, fhut, inclofeing, the chief or foremoft part; Ardent, on the inclofeing water part fide; Are, the earth and water, as Am is the furrounding earth; Area, the inclofeing water or mote ufually furrounding dwelling places; Arenofe, the inclofeing water place fide or fand; Arife, the lower fpring water; Arm, on the furrounding water or body part; Army, the body; Around, on the inclofeing fpring fide; Arow and Arrow, on the fpring; Arrive, on the river; Arfe, the fide, the lower or feat part; Art, on the parts or things; Article, the inclofeing water part fide fhuts, or joints; Artery, the water parts; Arithmetic or number, on the furrounding water parts; Artillery, on the water place fide.

Asaph N-Wales, the fea or lower water part. Afchaffenburgh on the lower Rhine, a burrow on the inclofeing water part. Afcot Bucks, the lower furrounding water fide or wood. Aferby Linc. the lower or fide water part or habitation. Afhburn Afhburnham, and Afhburton Der. Suff. Devon, Berks, on the lower or fide fpring part, vale or town. Afcalon a city of Paleftine, the lower inclofeing water bank, lane or ftreet. Afhdown, don or ton, Lanc. Yorkf. Eff. Oxon, the lower border or town. Afhele and Afhley Norf. the fide or lower water place. Afhelwell Thorpe Norf. the lower fpring water place gate part. Afheridge Bucks, the lower ridge, hill or high confines or edge. Afhford-fted-well and wood Kent, Surry, N-ham, Staff. the lower ford or way, ftation, fpring water or dwelling place and wood. Afia or Acia, the chief or firft country or cities. Afk and Afken Yorkf. a Saxon or the lower inclofeing water. Afkeadnith Scot. the lower water on the fide part. Afkerton Cumb. the lower confines water town. Aflakton Nort. the lower water place confines town. Afpley Gowiz or Guife Bedf. the lower or fide fpring water place part. Affenthire Scot. the inclofeing fide or lower er confines ſhire. Aſton, on the lower border or the lower town. Aſtalby Oxon, the lower ſide water place part or a-bode. Aſturia Spain, the lower water ſide or country. Aſt-well N-ham, the lower ſide ſpring water place. Hence As, the ſides or lower parts ; Aſcent, the lower water on the ſide ; Aſcertain, the lower water border ſide incloſeing; Aſh, the lower, oak being the higher; Aſhes, the lower ſide ; Aſhore, the ſurrounding water part ſide ; Aſide, the ſide or lower part ; Aſk, the lower water confines ; Aſlant or Aſlope, a low part on the ſide ; Aſleep, the low part ; Aſpen, the head low ; Aſpire, the low part high ; Aſs, the loweſt beaſt or below the ſide, as horſe is the higheſt, or the high on the ſide or ſeat ; Aſſay, the low up or in action or ſpring ; Aſſent, the lower within the ſide ; Aſſert, the lower water ſide ; Aſſeſs and Aſſize, the lower ſide ſet for the ſupport and defence of the border or country; Aſſign, within the border ; Aſſociate, at the lower confines water ſide ; Aſſuage, the lower ſpring edge; Aſſume, the low for the ſide ſpring; Aſſurance, on the incloſeing ſpring water ſide, where there is ſecurity ; Aſthma, the ſurrounding ſide low; Aſtern, the incloſeing water lower ſide ; Aſunder, the incloſeing ſpring water ſides.

Atcourt France, at the ſurrounding ſpring water ſide. Aterith Camb. Ire. the water ſide, way or ford. Aterton or Atherton War. Dur. the water ſide town. Athanaton, Thanet or Tanet Kent, N-Wales, on the lower ſide ſpring. Athel-ny, Atheny and Etheling Som. Norf. the incloſeing ſpring water place ſide. Athens, the incloſeing water ſide or below the water ſide. Athie, Athlon and Athol, Scot. Ire. the ſide ſpring, the ſide ſpring bank and the ſide ſpring water place. Atilburrow Norf. the ſide water place burrow. Atlas a moun-tain of Mauritania, on the lower water place or the tall. At-lantis iſland, at the lower water bank ſide or iſland, which according to the deſcription of it by the ancients, if it be not ſunk ſeems to be the weſt of England, which they ſuppoſed to be an iſland ſeparated from the reſt of this iſland. Atton or Auton Yorkſ. Scot. the ſpring ſide town. Atiſcoti a people of Britanny, at the lower water confines. Atticoti Scot. at the lower water confines ſide or wood. Atræ near the Euphrates, at the ſide water. Attrebati Berks, France, at the ſide water part ſide. Atſcar Yorkſ. at the lower border water or rock ſide. Hence Athirſt, at the lower water ſide ; Atheletic, at the incloſeing water place ſide ; Athwart, the ſpring water ſide which is acroſs the ſea ſide ; Atilt, at the water place ſide ; Atom, the all ſurrounding parts or bodies ; Atone, the border incloſeing; Atop, the covering part; Atrocity, the ſurround-ing

ing water part lower side; Attach and Attack, at the incloseing water side; Attain, the inside or coast; Attend, at the end or incloseing side; Attest, at the lower spring side; Attire, upon the side; Attitude, the spring part side; Attorney, on the border one; Attract and Attrition, at the confines water part.

AVALONIA or Glaston Som. on the spring water place side, bank or vale. Aubley and Aubery Wilts, the spring water place. Audry or Audre Camb. the spring water part. Avenmore Ire. the great incloseing spring or river. Alventon or Alvington Gloc. on the incloseing spring water place town. Averham Nott. the river vale. Aufon or Avon, an appellative of a river as appears under Aboyne. Augusta or London, the incloseing spring water side coast, street or lane town. Augustoritum in France, the incloseing spring border way or ford. Aukland Dur. the incloseing spring water bank side, land or water land. Aula, Aulau and Aulery of Tarsus, Germany and France, the spring water place side bank, haven or hall. Auldby Yorks. the spring water place side or an old part or habitation. Aularton Nott. a town on the spring water place. Aulre Som. the spring water place. Aulton, the spring water place or old town. Aulfric or Alfredswic of the Wiccians in Worc. the spring water confines or Wiccian water part side. Auntby Linc. the incloseing spring side part or habitation. Auseley War. the side spring water place. Ausoba Ire. the surrounding spring part or abode. Auteri Ire. the spring water sides. Aust, Austin or August, the spring lower side. Au or Aw, Awn and Awtanburgh Scot. Ire. Hunt the spring, the incloseing spring, and a burrow below the spring. Hence Avail, on the spring; Avant, the incloseing spring side; Auction, on the spring water side; Audacity, the spring water part confines or chiefest side; Audible, the side spring place of hearing; Average, the incloseing spring edge, or a payment made in Brittany and other parts of Gaul for supporting and defending the boundary rivers; Avenge, the incloseing water spring; Aver, on the spring; Aver-corn and Aver-penny, the spring corn or payment in lieu thereof for the lord of the soils proportion of the springs or fruits for the defence of the country; Averie and Avert, a low outside spring; Auf, of a spring; Auger, the incloseing water spring, which flows into the sea; Augment, the spring water on the side; Aunt, a spring or offspring of one side; Avoid, the surrounding spring side; Avow, the border spring; Author, the side spring man; Autumn, the surrounding spring side incloseing, decreasing, inning or

houseing

houfing the fruits of the earth; Awake, the fpringing action; Awkward, the water fide man or failors action or fpring.

Ax, Axey, Ex or Acç, Devon. Linc. the lower or fide wa-ter. Axelodunum or Hexham N-hum. the inclofing or high fide fpring water place town vale or home. Axantos or Ufhant, the inclofing water parts. Axholme or Axel-home, the furrounding fpring water place or ifland. Ax-minfter Devon, the inclofing water lower fide, ftation or monaftery Axona and Axmouth Devon and France, the in-clofing water and its mouth. Hence Ax or Ac-ç, the edge, the fide or lower water edge; Axiom, the furrounding or con-clufive fide; Axis, the round fide; Axel-tree, the round upon tree; Aye or Ay, the fpring; Azure, the leffer light fpring. Here we may obferve how little the letter x entered into the compofition of the Englifh language and their names of per-fons and places, which were framed long before the letter x or Saxon name was introduced amongft us by the Romans.

BABEL of Chaldea, the fpring water place part, habitation or dwelling place. Babingley Norf. inclofing fpring water confines part or dwelling place. Babham Berks, the vale dwelling part. Babthorp Yorkf. the fpring water or dwelling part gate. Hence Baby and Babble, a fpring part and the baby call, or on the baby.

BACONTHORP Norf. the inclofing fpring water part gate. Bacchus the fon of Jupiter by Semele, the fpring water part at the lower confines or meeting or kiffing the fea fide, or a colony of the Gauls grafted on the Titans or fea coafters. Hence Bacchanalian; Back and Bacon, the inclofing part; Backberond, bearing it within the precincts or limits, as the thief's having the venifon not only on his back but within the foreft and the extent of its laws.

BADE, Baden or Bode D-r. Wales, Germany, the inclof-ing or furrounding fpring water part fide, part or an abode. Badenoch Scot a part or an abode on the inclofing or confines fpring water fide. Badbury, Bodbury or Bedanburig Dorfet. the furrounding or border fpring water part fide, ftation or bur-row. Badeley and Badefley Chef War. the fpring water part fide or inhabited place. Badlifmere Wilts, Kent, an abode below the mere or confines water place fide. Badmanton or Badminfter Gloc. the inclofing fpring water part fide, abode or town. Hence Bad, Badge and Badger, the fpring part action at the meres and badgers place of refort; Bafile, the fpring off the water place or fea.

BAGINGTON War. a town on the inclofing fpring water part.

part. Bagmere Chef. the inclofing fpring water part lake. Hence Bag, the inclofing part; Baggage, an aged bag; Bagnio, the inclofing fpring water part inn.

BAINHAM Gloc. the inclofing fpring part vale or home. Baint, Baintbridge and Bainton Yorkf. Devon. the inclofing fpring water part fide, bridge or town. Hence Bail, on the boundary place; Bait, the fpring fide part; Bain, the fpring part within.

BAKEWELL Der. the inclofing fpring water dwelling place. Hence Bake, Baker and Bakehoufe, as the place of inclofing or putting water and meal together.

BALA, Baladeilin or Fale N-Wales, the fpring water place, vale or village, and a village at the lake fide. Bala Curi and Bala Monin Ire. a vale, dwelling place or village on the fpring or fea borders. Baldock Herts, the border water place or vale confines Balin Tobar Ire. the border part water place fide or bank. Balmerinock Scot. the fpring water vale or dwelling place meres on the fea. Balrodie, Baltarbet, Baltic and Balvenie Ire. Scot. Germany, a vale or village on the water part or boundary part, the upper water place, and the fpring water places. Hence Bailif and Bailiwic, on the village or boundary water place; Balance, the fpring water place fides or banks which are equal; Balcony, a part on the confines water place; Bald, the water place fide, which is moftly barren; Pale, the water part place; Ball, the all furrounding part; Ballad, the village fong; Balm, the furrounding vale place.

BAN, Bana, Bany or Bean Linc. Wales, Scot. Ire. the inclofing fpring water. Banatia Scot. the inclofing fpring fides. Banbury Oxon. the inclofing fpring part or burrow. Bangor, Banchor or Bonium N-Wales, on the inclofing fpring water part border, choir or monaftery. Banf Scot. the inclofing fpring water part. Bankyr or Banks Scot. the inclofing fpring water bank fide or border. Banna Venna, Banomana or Weedon on the ftreet N-ham Ire on the inclofing fpring water fide part, ftreet or town, it being the ftreet along which the Veneti, Wenta or Venedotæ made their migrations from eaft to weft, and thence the Irifh and Saxons called their country Ban, Ben, Van, Wen, or a part inclofed by water, as well as Totidan and Trivolac, a country on the border or covering water fide or furrounded by water. Banockburn and Banwell Scot. and Som. the inclofing fpring water or dwelling place. Hence Ban, Bane and Banifh, on, under or below the inclofing or boundary fpring part; Band and Bandy, the fide inclofing fpring part; Bang and Bank, the inclofing fpring part fides or action; Banner, on the inclofing fpring part or a river

vane;

vane; Banjan, the upper inclofure; Bannock, the fpring inclofing or baking bread; Banquet, at the fide of the confines
fpring water; Banter, the inclofing fpring water; Bantling,
the inclofing water place fpring part; Baptifm, the babe below
the furrounding water part.

BARAM Kent, the furrounding water. Barbacan Midd. the
water inclofing or bar fhutting part. Barbury-den-ford Wilts,
Linc. N-Wales, Yorkf. War. the fpring water part or bar burrow, vale and ford or way. Barking or Berking Eff. N-Wales,
the inclofing fpring water or bar part. Barkley, Som. Gloc.
Leic. the inclofing fpring water or bar place. Barkfhire, Ber-
cheria or Attrebati, the fpring water border or bar fide fhire.
Barnet Midd. the bank, hill or inclofing bar on the fide. Barn-
well Camb. N-ham. Ire. the inclofing fpring water part or bar
dwelling place. Barodon or Bariden Norf. Rut. N-ham. at
the furrounding fpring water part or bar fide. Barray or Bar-
rey Leic. Wales, Scot. on the water or bar part. Barfa, the
fea water part. Barton a parifh in W mor. on the fpring
water part or bar fide, farm or barton. Hence Bar, the water
part or fpring water part; Barb the bar part; Barbarous, the
bar part below or at its fide, as fhut like cattle or wild beafts
out of the community of rational beings; Barbel, the bar water place; Bard, the bar fide, who was originally the baron or
patriarch but afterwards became the Druids recorder; Bare
and Baren, on the water part, which is always bare; Barely,
the water part, bar or mere place; Bargain, the bar gain;
Barge and Bark, the fea bar or river veffel; Barley, on a barren place; Barm, covering or furrounding the fpring water
part; Barn, the fhut part; Barnacle, on the inclofing water
place confines; Baron and Barony, on the inclofing water part
or bar; Barrifter, below the bar fide; Barter on the water part
fide, where the fairs and markets were held Baronies feem
to derive their origin from the firft divifion of land by rivers
and banks, which muft be very ancient, and Barton feems to be
a leffer divifion of baronies into farms.

BAS Scot. an ifland, fide or a low water part. Bafepole
Ire. an ifland, low or fhallow water place. Baiham Norf.
the water part fide vale or home. Bafingftoke Hants. the lower
confines water part ftation or feat. Bafingwerk N-Wales, the
lower fpring water confines. Paffe or Bleffium Heref. the fide
or low fpring water place. Baffenrig Scot. a low part at the
fea fide. Baffitania Spain, the coaft on the lower water part
fide. Hence Bafe and Bafhful, a low part; Bafhaw, the fpring
fide part; Bafis and Bafs, the lower part; Baftard, a bafe fide

fpring

fpring part; Baſle, the lower ſide part; Baſtion, on the lower ſide part.

BATAVIA, Batavodorum or Holland, the ſide ſpring water part way, ſtreet or gate on the high ſurrounding ſpring water place bank ſide or land. Baterſea Surry, the ſide ſpring water part or iſland. Bath, Batherton, Bathgate and Bathſtone Som. Chef. Scot. the ſide ſpring water part ſide gate or town. Hence Bat, Bate, Batement, Bath, Bating, Battery, and Battle, the ſide ſpring part.

BAVARIA or Boiaria in High Germany, on the ſurrounding ſpring water part or the country of the Boii, a Gauliſh nation ſettled on the Danube. Bavoid caſtle Kent, the ſpring water part ſide caſtle. Bauli Italy, the ſpring water place. Bawdley haven Suff the ſpring water part at the ſea or an haven. Baw or Bew caſtle Scot. the water ſprings or quick ſands caſtle. Hence Bavin, an incloſed ſpring part of wood; Bawble, the water place ſpring part, bubble or water bladder; Bawd, the ſide ſpring part of the game; Bawl, the high or call ſpring part; Bay, the ſpring water part, a ſpring part or ſea at the ſpring part and its colour.

BEACHY Suff. the ſpring water part confines. Beamfleet Eff. the ſurrounding ſpring water part flood. Beaumares N-Wales, the ſea ſide ſpring water parts or quick ſand marſh. Beaulieu Hants, the wood ſpring part on the river place. Beaumannor park Leic. the manor or border ſpring water part. Beaudley or Beawdley Worc. the ſpring water part ſide place. Debba N-hum. the ebbing water part. Hence Be, the ſpring part of things or life; Beach and Beak. the ſpring water part confines; Beacon, on the beach; Beadle, on the ſpring water ſide; Beam, the ſurrounding ſpring parts of light, water, ſcales or ſpringing timber; Bean, an incloſed or ſhut ſpring part; Bear, on the ſpring; Beard, on the ſide ſprings, Beaſt, the lower ſpring of animals; Beat, the ſpring part at the ſea or ſide; Beau, a ſpring part man; Beaver, the ſpring water part; Beauty, the ſpring or ſpring water ſide part

BEC, Beck or Beg, the confines ſpring water part. Becket-hill, the confines ſpring water part ſide or hill. Beckford, the ſpring water part ford or way. Beckenſgill Lane. the confines ſpring ſide water place. Hence Become, the ſurrounding water confines or common ſpring part; Becauſe, by or to be the confines water part uſe; Bechance, by chance; Beck, Becken and Beg, from beak and beacon; Begin, the ſpring acting in; Beget, the ſpring part at the water confines or ſea ſide.

BEDAL Yorkſ. on the ſpring water place ſide. Bede a ve-nerable Engliſh hiſtorian, who flouriſhed in 694, the ſpring

part fide, an abode or abbot. Bederic and Betorix, the king's
abode. Beddington Surry, the fide fpring part or bedding
town. Bedford, Lettidur or Lifwider, the fpring fide part
way or ford, place or palace. Bediford Devon. the fpring part
fide ford way or abode. Beddingfield Suff the inclofing fpring
water part fide dwelling place. Bedwin Wilts, the inclofing
fpring water part fide, abode or bed. Hence Bed, the fpring
part fide or an abode; Bedawb or Bedaggle, the bed or fpring
part at the confines water place; Bedew, the fpring part fide
water.

Bees Cumb. the fpring water part fide, cell or dwelling
place. Beechworth Surry, the fide fpring water part confines or
by the beech or fpring water part bank. Hence Bees, the fmall
inclofing fpring part or beings; Beech, the fpring confines
parts; Beet, the bee at or at the fpring part; Beeves or Beuves,
the fpring part fide parts or animals; Befall, the fpring part
on the water place; Befit, the fpring part pit; Before. the
fpring part on the furrounding water part; Befriend, the fpring
water part on the fide; Behalf, the fpring part half; Behave,
the fpring part haven or fhelter; Behead, the fpring part head;
Behind, to be on the fpring on the fide; Behold, to be on the
high water place fide; Behoof, of the fpring parts or profits;
Being, fpringing or living.

Belishanon Ire. on the lower inclofing fpring water place.
Belerium, Antivæfteum or Helenum, the utmoft promontory of Cornwal, on the fartheft or the fide water place. Be-
lefme, Belifhama or Bhibel Lanc. the loweft or fartheft furrounding water place or river. Belgæ, Belke or Morini of
France, Hants, Wilts, Som. Ire. the water place or fartheft
water confines on the inclofing water vale or low country, the
Belgæ being a branch of the Cumbri Galli or dwellers in vales
and the radical b in bale or bele by its inflecting power changing into f and v, fo as to make fale or vale. Belin or Caffi
Belin chieftain or general of the Iceni, the inclofing or Iceni
water place border w rden. Bell-ifle, the fartheft or loweft
water place ifland Bellæ Italy, the fartheft place or water
place Bellaife or Belleland Dur. the loweft or fartheft water
place bank fide, vale or land. Bell defert War. the water or
fartheft place defert or wood. Belleri or Bellers War. on the
water or fartheft place or vale. Bellatucadrus W-mor. the fartheft water place fide or vale keeper. Bellafitum Oxon, the
water place or fartheft fide. Belfey N-hum. the water place
fide. Belfars hill Camb. the water place or fartheft fide hill,
which, as well as various other names of bal and bel ferve to
convince me that the Belgæ were originally Gauls, and generally

rally concerned in the firft plantation of this ifland, and were perhaps the very Cambels, Cimbels of the Ifceni or Caledonians who fought Julius Cæfar. Hence Belabor, the fpring part on the furrounding water part or at the river's mouth; Belated, the fpring part on the water place fide; Belay, a bye place; Belch, fpring up action; Belie, to be or a fpring from light; Belief, fpringing from light; Bell, a high calling fpring part; Belley, on the food place, or bolly, the ball place; Belong, the inclofing or confines water place part; Beloved, the furrounding fpring water place fide part; Below, the furrounding fpring water place; Belt, the fide water place; Bemoan, to be on the furrounding water or fea.

BENEFICA Herts, the inclofing fpring water parts. Benbridge, Benenden, Bengeley and Bengworth Kent, Worc. Salop, Yorkf. N-hum. the inclofing fpring water part edge or bridge, vale or den, water confines place, cell or fpring water part. Benones Leic. on the inclofing fpring fides. Benington, Benfington and Benfon, a part or town on the inclofing fpring water fide. Benno or Beuno N-Wales, on the fpring water part. Benfbury, Bensford, Benftead and Bently Surry, N-ham. Herts, Suff. the inclofing fpring part fide dwelling place or burrow, ford or way, ftation or feat and place or pa-
·. Beneventum and Maleventum Italy, on the inclofing
...   ... and on the furrounding water place or
          ...   ... mans meant good and bad water
places, as appe.   ... the radical bene or ben in the
Latin tonsue, b.. the Fr.... frith and old Britons ufe ban
or ben in a more priv.. fen....   ... ence Bench, on the fpring confines part or bank; ... or ... a fpring part on the fide, Beneath, the inclofing ... ing part of the water fide; Benefit, Benign and Benifon, the fpring in on the fide or in the pit; Benumb, the fpring part without the furrounding fpring part.

BERCARII or Vaccarii Yorkf. the water or fpring water borders. Berdiey N-Wales, the ... water part ifland. Bere, Berew, Bery or Turville Devon, Dorfet, Dur. N-Wales, the fpring water part or place. Beresford, Berkhill, Berkhamfted, and Bermingham Kent, N-ham. Herts War. the fpring water part ford or way, hill, vale or home ftation and water confines vale or dwelling place. Bergos or Berg.e, the furrounding or fea water part. Bernard Caftle Dur. on the inclofing or fpring water part fide caftle. Bernicia, on the inclofing fpring water part upper confines, as deira, is this fide of the divifion water or Tyne; the former kingdom in the time of the Saxons exten-tending from the Tyne to the Frith of Edinburgh and the
latter

latter from the Tyne to the Humber. Beridale Scot. the spring
water part vale. Bernswale Scot. the inclosing spring water
part vale. Berigonium or Bargeny Scot. on the spring water
part confines. Berybank Staff. the river bank. Beryfield Bucks,
the river dwelling place. Berselin Staff. the river side bank.
Berstaple Devon, the river station place Bertha or Perth Scot.
the river side part or grove. Berton Staff. the river side town.
Berubium or Beru Scot. N-Wales, the spring water part or
abode. Berwick, Borcovicus or Aberwic N hum. the spring
water part confines, street, fortification or harbour; and in
ancient charters Berwic was a law term, by which the privileges of the border waters belonging to townships were conveyed or granted. Bery Pomery Devon, the spring water part
on the surrounding water or sea part. Hence Berry, on the
water part, as belonging to the Britons, as apples did to the
dwellers in vales or the Gauls from whence they were denominated by the Gauls a vale or a fale. the inflection of apple.

BESILE Hants, the spring water side or lower place. Beson
Spain, the surrounding spring side or lower part. Besbicus an
island of Bithynia, the lower or island water part. Hence Beside, the spring part or by the side; Besiege, the spring part
edge; Besmear, the spring side surrounding water part; Be-
sot, the surrounding spring part side; Best, the spring part
lower side; Bestir, the spring part lower side water.

BETHAM W-mor. the spring part or river side vale or home.
Bethlem Midd. on the surrounding spring water place side.
Bethesley Gloc. the side spring part lower place. Bethmesley
Yorks. the side spring part middle water place. Betorix, Be-
terice or St. Edmundsbury Suff. the upper border spring part.
Betus N-Wales, the spring part side, house or abode. Hence
Bet, the side spring, Bethral, on the side spring water part;
Betray, from the side spring part; Better, the side spring part
water; Between and Betwixt the water within the spring part
sides.

BEU or Bueth castle Cumb. the spring part side castle.
Beverey Worc. the confines spring water part. Beverley
Yorks. the spring water part place. Beverston Gloc. the spring
water part side town. Bevers castle Leic. the spring water
side castle. Bevis tower Surr. the river side tower. Bewdley
Worc. the spring part side or dwelling place. Beward, an exchange of wares betwixt the Saxons and Britons, on the spring
water part side. Hence Bevil, the spring water place; Be-
verage, the spring water part edge; Bevy, the spring parts or
beings; Bewail, the spring part water waves; Beware, on the
spring water part; Bewitch, the spring confines part edge; Be-
way

wrav, from the spring part; Beyond, on the surrounding spring part side.

BIBRACTE, Bibroci or Berks, the border spring water confines dwelling part. Hence Bib, the spring dwelling part.

BICESTER or Bisister Oxon, the spring dwelling city Bicknor Heref. a dwelling part on the confines water border. Hence Bicker, the confines water dwelling; Bid, the spring part at the side.

BIGIN or Wigan Lanc. a dwelling on the spring head or first part. Biglefwade Bedf. the side spring confines water place dwelling. Hence Big, the confines water dwelling part , Bigamy, two marriages or surrounding water parts; Bigot, the surrounding water or sea confines dwelling.

BILDAS or Bildewas Salop, the spring water part side dwelling place. Billangho Lanc. the high spring water bank dwelling place. Billefdun N-hum. a dwelling place on the hill side. Billing N-hum. the inclosing spring water dwelling part. Billingfgate, Billingfbere, Billingfton and Billington London, Berks, Yorkf. Nott. the inclosing spring water dwelling part gate, or town. Billericay Effex, a dwelling part on the border. Hence Bile, the spring water place of life; Bilge, the spring water place edge or side; Bilk, from the dwelling place; Bill, on the dwelling place.

BINCHESTER or Binovium Dur. the inclosing spring part or dwelling city. Bindon Dorfet, the inclosing spring part or dwelling town. Bins Scot. the inclosing spring or dwelling part. Hence Bin, the food inclosing part; Binary, the spring water inclosing sides; Bind, the side inclosing spring; Biography, life writing.

BIRCHANIS or Birchynis, the inclosing water dwelling **part** island. Birdlip Gloc. the spring water part side high place. Birklec W-mor. the inclosing water brook dwelling part. Birling Kent, the inclosing water place dwelling part. Hence Birchen, on the spring water confines part; Bird, the high animal springs or things; Birth, the springing into life or the poffeffed parts.

BISCAW or Bifcaw Woune Corn. a part or habitation on the furrounding or weftern water fide. Bifham Berks, the spring water part side vale or home. Biffemed Bedf the furrounding foring part fide. Bifus, the spring part fide or dwelling. Hence Bifect, the spring fides act of dividing; Bifhop, the inclofing fpring water fide, living or dwelling part.

BITFORD, Bitham, Pithefden, Bittern and Bitton War. Linc. Bucks, Hants, Gloc. the spring part fide or dwelling, way or ford, vale or den, on the water part or town. Bivcll
N-hum.

N-hum. the ſpring water dwelling place. Bixbrond, Oxon, the ſpring water part confines on the hill ſide. Hence Bit, Bite, Bitter and Bittern, the ſide ſpring water part, which ſeparates and mixes with the ſea.

BLACKAMORE, Blackburn, Blacklow, Blackmere, Black Mountain, Blackneſs, Blackney and Black-tail-point Yorkſ. Dorſet, Lanc. War. Salop, S Wales, Scot. France, Norf. Eſſex, the incloſing water, ſpring water, low or black moore, neighbourhood, vale, mere or boundary, mountain, promontory, confines or coaſt point Bladin hills Ire. the incloſing ſpring water place ſide hills. Blaen Leveny N-Wales, the incloſing ſpring water place. Blandford Dorſet, the incloſing ſpring ſide ford or way. Blank Caſtle S-Wales, the incloſing ſpring water place confines or water bank caſtle. Blankeney or Blankveney Linc Gloc a part on the incloſing ſpring water place Blath Staff the ſide ſpring place. Blatherwic N-ham. the ſide ſpring water place confines, ſtreet or fortified part. Blatum, Bulǝium or Bulaeſs Cumb. the ſide water place bulge, jut or promontory. Hence Black, a ſhut or low place; Blackmaile, the middle or border ſpring water place confines or a payment made to the border keeper in Dur. N-hum. Cumb. and W-mor; Bladder, the ſide ſpring water place part; Blade, the ſpring place ſide; Blanch and blank, on the ſpring water place confines or bank; Blaſt the ſpring water place lower ſide; Blaze, the ſide ſpring water place.

BLECHINGDON Oxon. the incloſing ſpring water place town. Blencarn, Blencow and Blankenop W mor and Cumb. on the incloſing ſpring water place confines, bank or hill and ſouth part. Bleſtium, Hean or old town Heref. the ſide ſpring water place, the incloſing ſpring water place or old town. Bletſo or Bleaſho Bedf. the ſurrounding ſpring water place ſide. Blethen Cloith, Caer Palladur or Bath. Som. on the ſide ſpring water place ſhut or the city of Pallas or Minerva a tutelar deity of the ſprings. As to the ancient tradition that thoſe ſprings were diſcovered by a Britiſh king of this name, it muſt allude to Hercules the cleanſer of the Augean ſtable by drawing thro' it the river Alpheus, whoſe image was found on the walls there. Hence Bleach, Bleak, Blear and Bleat, the ſpring water place confines; Blemiſh, below the ſurrounding ſpring water place; Blend, the ſpring water place in the ſide; Bleſs, the ſpring flowing on the lower place; Blew, the water place ſpring.

BLISWORTH N-ham. by the flowing ſide water ſpring. Blith Suff. Nott War Staff. N-hum the ſide flowing ſpring part. Blidfield and Blith Burrow Staff. and Suff. the ſide

E flowing

flowing spring part dwelling place or burrow. Hence Blind, without the internal spring of light; Blifs, the light spring flowing on the lower parts; Blister, the water part below or at the side of the spring water place; Blith, the side flowing spring part.

BLOODGATE near Brancaster Norf. the furrounding water place side or flood gate or way. As to the herb ebulum found here, and suppofed to have been called Danes blood from fome great flaughter of the Danes in this place, it, like various other conjectures founded merely on a wrong derivation of names, feems to be a very great miftake; for Dan's blood in the Anglo Britifh language fignifies at the flood fide. Moreheath Staff. a heath on the furrounding spring water place. Hence Bloat, the furrounding spring water place fide; Block and Blockade, the furrounding spring water place fhut; Blood, the side furrounding spring water place; Bloom, Blow and Bloffom, the flowing spring water place around; Blot, the furrounding water place or fpot.

BLUNDEL and Blunt, the inclofing spring water place fide or vale. Hence Blubber, fpringing up the water part; Blue, the spring of light part; Bluff, the spring water place huff; Blunder, the spring water place under; Blunt the spring water place unto; Blur, the spring water place; Blufh, the spring water place low; Bluster, on the spring water place fide part.

BOADICEA or Bonduca queen of the Ifceni, the Kentifh fide or Ifceni furrounding water part or habitation. Bocking, Kent, the Kentifh inclofing water confines. Hence Bo, the circle or furrounding water part; Boar, on the furrounding water part; Board, on the furrounding water part fide; Boaft, at the furrounding water part fide; Boat, at the furrounding water part; Bob, the furrounding water part spring part.

BOD or Bodiam Suff. Oxon. Gloc. Wales, the furrounding water part fide, abode or ftation. Bodincomagus Italy, the furrounding water part in the vale. Bodionti in France, on the furrounding water part fide or vale. Bodotria Scot. the furrounding water part fide border port or haven. Bodmin Corn. the furrounding water part edge or the mining habitation. Bodowen, Boduacus, Bodowir, Bodvari and Boduni or Wicci, Wales, France, Oxon. Gloc. the furrounding spring water part, fea or marfh fide, abode or ftation. Bogchilt Scot. the furrounding water confines at the hill fide. Hence Body, the furrounding part either of man or a country; Bog, the furrounding water part confines; Boil, the furrounding water part hot flood.

BOLD

BOLD Lanc. on the furrounding water part fide. Bolebec Bucks, on the furrounding water place fpring water part. Bo en or Bo'in France, Chef. the furrounding water part or fea bank Bo teby and Bolton N-hum. Yorkf. and Bold and Bolt, on the furrounding water part fide and town.

BOMARII Italy, on the fea furrounding water parts. Bohemia, the high furrounding water part edge or fide. Hence Bomb, the furrounding water part fpring part, as a river difcharging into the fea.

BONBURY or Wodenfburg Wilts, a dwelling or burrow on the furrounding water part. Bongey Suff. the confines on the furrounding water part. Bonville or Bonwell Heref. a fpring water dwelling place on the furrounding water part. Bonium or Bangor and Bonn N-Wales, Germany, on the furrounding water part. Hence Bond, the furrounding or inclofing water part on the fide; Bone, in the furrounding part; Bonny, a fpring in the furrounding parts; Bonfire, a fire in the furrounding parts.

BORDAR Corn. on the furrounding water part or fea fide or border. Bordarii, the borderers, or thofe who defended them under the borrow nolder or head burrow. Boreum or Venicnium, the fpring on the furrounding water or fea part or a haven. Borefionefs Scot. a town on the lower furrounding water part or border. Borth and Borthwic N-Wales, Scot. on the furrounding water part fide, port, ftreet or fortification. Borwic or Borovicus N-hum. a ftreet, fortification or haven on the fea fide at the river mouth. Near here at Pont Eland, or the furrounding fpring part bank fide or land, the Romans according to their ufual practice of placing thofe of the fouthern parts over the north and thofe of the north over the fouth parts, ftationed a cohort of the Cornavii, as they did the Dalmatian horfe which had been raifed here amongft the Tyne Meatæ, at Bruncatter in Norfolk under the count of the Saxon fhore, and from whom the deles, vale dwellers or Cumberi of England, Scotland and Wales in a great meafure derive their origin; the notion of the Romans drawing thefe levies from the continent being entirely groundlefs, fince the people appear to be the original planters of Britain. Hence Book, Boom, Boot and Both, as being parts furrounding or covering; Border, the furrounding water part fide; Bore, the fpring to the furrounding water part or foot; Boreal, the upper border; Born, in the border; Borough, the border fpring confines or dwelling place.

BOSCHATS Scot. the furrounding water part or fea fide confines or inclofing part. Bofeham, Bofton, Botherfon,

Bot-

Bosphorus, Bostock and Bosworth Suff. Linc. Wales, Chef. Leic. the surrounding water part or sea side vale, town, lower way, station and side spring water or river. Hence boscage, Bosom and Boson, as being on or about the sea side.

BOTHWELL Scot. the surrounding water part side spring or dwelling place. Bothal castle N-hum. or Glanaventa, a castle on the surrounding water side or the spring water bank, or coasts; where the Romans placed a garrison of the Morini, Veneti or Venta Belgarum. Botolph or Botle a town seated on both sides the river Witham in Norfolk, and Bottlebridge Hunt. a part, dwelling or bridge on the surrounding water part sides, or an abode. Botontines N-ham. an abode on or the border or inclosing water side. Hence Botany, on the surrounding water part side; Botch, the surrounding water part side edge; Both, the surrounding water part sides; Bottle, on the surrounding water part side or the abode place; Bottom, about the surrounding water part side.

BOVERTON, Bovium or Cowbridge S-Wales, the spring surrounding water part town or inclosing spring water part edge or bridge. Bovinda, Boandus, Buvinda or Boyne Ire. a part or country on the surrounding spring water edge or side. Bowes and Bowland Yorkf. Lanc. the surrounding spring water part side, bank and bankside or land. Boxhill and Boxley Surry, Kent, the surrounding water part hill or place. Boyle Ire. the surrounding water part place. Hence Bouge, Bough, Bounce, Bound and Bounty, as springing beyond the surrounding or body part; Bouse, the surrounding spring side; Bow, the surrounding spring part; Bowels, below the surrounding part or body; Bower and Bowl, on the surrounding spring part; Box, the surrounding water confines part or man; Boy, the surrounding spring water part.

BRABANT a dutchy of the Low Countries, the water part side, a bottom or low country. Brachium or Brachy and Brachley Yorkf. S-Wales, N-ham. the water part confines or inclosing place. Bradburn Der. the inclosing spring water side or country. Bradenham Bucks, a vale below the water part side. Bradenston or Britendun Wilts, a town below the side water part. Bradford-gate-ing-ley-ston and Wardin, Wilts, Yorkf. Leic. Hants, Gloc. S-Wales, the water part side or side water part ford, gate, confines, place, town inclosing spring side. Brae of Mar Scot. the country of Mar, or the surrounding water or sea side. Brage, Brigg or Broughton Hants, Yorkf. the inclosing water confines parts or countries Braibrooke N-ham. the water part, rotten or broken brooke. Braich y ddinas Wales, the part above the city.

Braid

Braid Alben Scot. the country along or at the fide of the hilly ends. Brakenbee and Brakenfey Yorkf. Scot the brooke and fea confines water part. Bramhope, Bramifh and Bramton Yorkf. N-hum. the furrounding water part fpring part, fide or town. Bran, Bren, Breany or Bryn Norf. N-Wales, Ire. on the inclofing water part, hill, cell, baron, judge or king. Brancafter or Branodunum the inclofing water part hill, fortification or city. Brandenburgh in Germany, a burrow below the inclofing water part. Brandon Salop, on or above the inclofing water part fide. Brandrith or Galaber W-mor. the inclofing or fpring water place ford or haven. Branogenium or Worcefter, the inclofing water part confines or hill. Branfford on Watling-ftreet Leic. the inclofing water part fide way or ford. Branfpeth caftle Dur. a caftle on the inclofing water or hill part. Branten, Branxton and Braunton Salop, Norf. Devon. a town below the inclofing water part. Bratton, Suff. a town at the water part fide. Brawerdin Devon, on the fpring water part fide. Bray or Brey Ire. Berks, the fpring water part. Hence Brace, the water part fhut; Brackifh, the fpring water part lower confines; Brad, the water part fide; Brag, the water part confines; Braid, the fide water part; Brain, the infide or internal part or water part; Brake and Bramble, the water confines or furrounding parts; Bran, the inclofing part; Branch, the inclofing water part confines or divifions; Brand, the inclofing water part or fide mere, which is a mark of the boundary; Brafs, the leffer water part; Brave, the fpring water part or part; brawl, on the fpring water part; Brawn, the animal inclofing or fide part; Bray, the fpring water part.

BRECHIN and Brechinoe or Loventium Scot. S-Wales, the inclofing fpring water part confines or the Veneti or Venedotian furrounding water place confines or the Venta Silurum. Breede and Breeden Suff. Wilts, the fpring water part fide, den or vale. Breich Scot. the upper fpring water part. Brember Suff. the furrounding fpring water part. Bremenium or Brampton N-hum. the furrounding fpring water part town. Bremetonacum or Overburrow Lanc. the furrounding fpring water town or burrow. Bredanicum Mare, the Britifh fea, and Bredan or Brydan feems to be the proper name of Britain. Bremen in Germany, on the furrounding fpring water part. Bremicham or Birmingham War. the furrounding fpring water part confines vale or dwelling place. Brendam Scot. Ire. below the inclofing fpring water part. Brennus king of the Veneti of Gaul, who conquered many nations and ranfacked Rome, a judge, chieftan or king from his guarding

the

the confines and determining difputes on a hill. Brent, Brent-
knoll-marfh tor-wood Eff. Midd. Som. Devon, the fpring water part on the fide, border hill, marfh, town or wood. Bre-
flaw Germany, the fpring water part fide place. Bretagn,
Britain, Bredan or Brydan, the inclofing lower or fide water
part coaft. Bretenham or Combretonium Suff. a vale below
the water part fide. Breton Italy, on the water part fide.
Betwell or Bretevil Berks, the water part fide fpring or dwelling place. Breulnis caftle Gloc. the fpring water place fide
caftle. Breva or Brivæ, the fpring water parts. Brewood
Staff. the fpring water part fide or wood. Hence Breach,
Breech and Break, the confines or inclofing water part;
Bread, at the water part; Breadth, the water part at the fides
or from fide to fide; Breaft, at the water part fide; Breath or
Breeze, at the water part fide or wind; Breed, at the fpring
part; Breviate, the fide fpring water parts; Brew, the fpring
water part.

BRIAREUS a Titan fo called by the gods but amongft men
Ægeon, the lower country fpring water parts or fides. Bride-
kirk and Bridfalle Cumb. Yorkf. the fide fpring water part
together with the fea border and the fpring water lower part.
Bridge-ford-north-water Surry, Salop, Som. the ford, north
and fpring water part, confines or bridge. Bridlington and
Bridport Yorkf. Dorfet, the confines or fide water place town
or port. Brigantes, Birgantes or Blani of Deira or Dur.
Yorkf. W-mor. Lanc. Cumb. and of Ponticu in France, Spain
and Ireland, the inclofing water part or Kentifh confines or
coafts, who feem to be a mixture of Cumberi and other Britons. Brigantium or York, the inclofing water part fide or
city of the Brigantes. Brigcafterton, Brightftow and Bright-
helmfted Rut. Gloc. Suff. the confines water part fortified
town, ftation and fea ftation. Brill or Bury hill Bucks, the
fpring water part place or hill. Brim, the water part rim or
edge. Brinklow War. the fpring water place brink or edge.
Brinleton Som. a town on the lower water part fide. Briftol
or Briftow Som. the lower water border or ftation place. Bri-
tanhuis, faid to be a watch tower built by the emperor Clau-
dius on the fea coaft of Gaul, the fide fpring houfe or dwelling place below the water part fide. Brittain, Beritan, Bretan, Bredan or Brydan, the inclofing, lower or the fide water
part coaft or country or on the lower or water part fide or coaft.
A name given to this ifland, or that part of it which lies oppofite to the French coaft, by its firft Celtic planters, after their
ufual manner of naming countries by the nature of the fituation, as appears by fuch of them as are here defined. As a

specimen of our being grofsly mifled from the truth by our
antiquaries and etymologifts, we prefume here to fubmit Mr.
Camden's explication of the name of our country to the confideration of the public. He labours hard to derive the name of
our ifland from its painted inhabitants, and endeavours to fhew
that all countries were denominated by the names of its inhabitants; as Media, Perfia, Saxony, England, Scotland, Bel-
gium and fome others, and that the Britons were accuftomed
to paint their bodies; for fays he, almoft all the ancient Britifh names partook fomewhat of a colour, as Cogidunus, Argentacoxus and Segonax of the red; Mandubratius, Cartifmandua, Togodumnus and Bonduca of the black; Venutius
and Imanuentius of the white; Vellocatus, Carvillus and
Suella of the waterifh colour; Cuniglafus of the blue; Cungetorix and Arviragus of the golden colour; Prafutagus and Ca-
ractacus from a lively or brifk colour; and Adminius from the
vermilion, befides other current Britifh names, as, Gwin,
white; Coch, red; Luid, brown; and Du, black. The ancient Britons, fays he, who were thus accuftomed to paint their
bodies, in order to diftinguifh themfelves from their anceftors
the Gauls, named their country Brith-tan, the country of the
painted men, as other nations were accuftomed to do, according to their natures, inclinations, cuftoms and employments;
as, the Ionians from Javan their founder, the Ifraelites from
Ifrael, the Chananites from Chanan the fon of Cham, the
Iberi as being miners according to the Hebrew derivation,
the Heneti as being wanderers, the Nomades as being bufied
about cattle, the Germans as being warlike, the Franks as
they were free, the Pannonians from their pannas or long fleeve
cloth coats, the Æthiopians from their blackness, and the
Albans from their white hair. And that the Greeks, who
traded here for tin, added tania to Brith, as being expreffive of
a region, as in Mauritania, Lufitania, Turditania and Bafti-
tania, the country of the Mauri, Lufi, Turdi and Bafti, like
Denmark, a compound of the Danifh and the German word
March, a boundary or limit. Our author's conjectures being
thus wholly grounded on the derivations of names, our referring the reader to their explications in their proper places
might fuffice for his confutation, fince they clearly difprove all
that he has advanced, and fhew that countries and places were
firft named after their nature, circumftances and fituation,
and thence nations and perfons; and that this original name of
Britain, as well as thofe of other countries, was wholly impofed by its own original planters in their own dialect, though
intruders have been fometimes accuftomed to vary or attribute

the original into their own dialects, as was the case with Denmark, which at first in the time of the Cimbri went by the name of Cimbrica Cherfonefus, the Chimbrian promontory; afterwards Wetallaheed, a compound of English and Welsh, signifying water all along; since Jutland or Gutland, the water confines or Goth possessions; and now Dania or Tania, the inclosing or lower water part, of the same signification as Tan in Britain and not a region; and Brith is a russet colour or an intire natural mixture of red and white with black, different from the colour of glastum, and neither of which could be a very proper colour for painting on some of the ancient British canvass, but all that can be possibly imagined in this respect seems to be, that the sandy Picts of Scotland, to whom the custom of painting has been chiefly imputed by the general run of historians, and whom the Romans considered as a different nation from the British, might, in order to appear the more amiable to some of the southern inhabitants, stain their skins or skin garments with glastum. It may be also insisted from Polybius, Cæsar, Diodorus, Tacitus and others, that a considerable part of Armorican Gaul went by the name of Brittany in their days, without any mention made of their painting their bodies; and that Cæsar, the first broacher of this story as well as invader of the country, who could have but a very imperfect notion of the customs, manners or languages of the Britons, must have been misled as well as Mr. Camden, by mistaking the meaning of those supposed names of colours; or maliciously forged the story with a view of palliating his unjust invasion of the country, like many other Roman forgeries as to the customs and manners of the chief Celtic nation, very unlike their other customs; as, that the Britons, instead of animal compensations, sacrificed in their wicker basket, innocent men and not beasts, or delinquents against the Druid religion; that they were accustomed to promiscuous concubinage, as having in those days of simplicity no apartments in their houses, any more than many of the present Britons, who are not therefore to be charged with so heinous a crime; that they went naked, as they usually stripped, like the Gauls, when they engaged in battle or were employed at their daily labour; or that they were cloathed only in the skins of beasts, though their own historians inform us that the most dignified wore fine long garments, girdles, and golden rings, bracelets and other ornaments; and that they fed on human flesh, when they abstained from that of animals, and as they professed Druidism, must have been prohibited the use of animal food. But if we compare those accounts with the principles

ciples and practice of the Druids, in other refpects, as their notions in phyfics and meta hyfics, the immortality and tranfmigration of fpirits, in a future ftate of rewards and punifhments, aftronomy, geogra hy, languages, and various other branches of the arts and fciences, we can give but little credit to thofe traducers of the Britifh antiquities and invaders of the country. Befides, as the Roman hiftorians confidered the Britons and Picts as diftinct nations, and they charge only the Picts with the cuftom of painting, it feems more probable that our ifland would have been named Pictland than Britain, or Britland or the painted country, but fo far is it from being true that countries have been named from the complexion of the inhabitants, or their cuftom of painting, or any other cuftoms, that thofe cuftoms and qualities all borrow their names from thofe of the countries, as appears by their definitions in this work.

It may not perhaps be improper here once for all to obferve, with refpect to the inveftigation of the origin of nations from their names of perfons and places, that all names being originally appeliative, and ferving to defcribe rather than to denominate a people, they as fuch can afford no fort of affiftance therein, but as diftinct generical names or radicals, invariably made ufe of by different clans and nations, and employed as the models and ground wo k of their refpective dialects, they ferve to defc ibe and diftinguifh the feveral nations and tribes of every colony and country. Hence, for inftance, we are furnifhed with fufficient traces of the Titan, Phrygian, and Britifh migrations along the Mediterranean, Spanifh, Armorican, and German fea coafts. And of their being thofe whom Bochart and fome others fuppofed to be Greeks and Phœnicians; and for difcovering that they made their migrations through thofe feveral countries into Britain not long after the deluge; leaving colonies in their way for plantation and cultivation on the fea banks, promontories, and tops of mountains, in the days of Jupiter, Pluto, and Mercury, before the vales, flats, and midland countries were drained by rivers, cuts, and banks, and had recovered the ill effects of the deluge and were fit for cultivation; fo as to enable any to inhabit the vales, plains, villages, and cities, or to fucceed in their fucceffive inland migrations; as appears by the ridges, trees, flats and inclofures on the tops of mountains, and trees, bogs, fhells and fifh bones in midland parts of the earth. In the fame manner we may fee this firft weftward migration fucceeded by another of the Celtes, Iberi, Celtiberi, Silures, Guithels, Caledonians, Gerions, Greeks, and Germans, the planters of

F   groves,

groves, trees and vineyards, that the firſt inland migrations were of the Commatas or dwellers in vales or villages, who, retaining the ſame vicinity, pierced forwards in ſtreets or alleys along the ſide of rivers, which they had made, and were thence named Gallia Commata or the Gauliſh Connaots or Cantons; and that all thoſe on their arrival in ancient Celtica coaleſced with the Briges, Iberi and Britons of the ſea ſide, as is alluded to in Diodorus by the marriage of Hercules with the daughter of Aleſia, and begetting on her Galates, the founder of the Gauls; and from this union of landmen and ſeamen or dwellers in vales, and ſea coaſters they were called Cumberi and Gomer Gallus; as were the Celtiberi, from an union of the Celtes and Iberi, before Hercules had ex elled the cattle of Gerion or the Greeks and Germans, the weſtern water confines, and Atlantics out of Spain into Germany, Britain, and the Atlantic coaſts, according to Diodorus and Plato's Timæus. Here, as well as in other countries, the Herculean colonies or Cumbri employed themſelves in cleanſing the Augæan ſtable, or draining countries by drawing rivers through them, deſtroying wild beaſts, civilizing mankind, and erecting cities, according to what ſeems to be intimated to us by the Herculean labours; as were the Teutons, another branch of thoſe people on the banks of the Rhine; and the Dæ, Dacæ or Getæ, the deſcendants of Tyras and Rhip hath on the ſide of the Danube. And that all thoſe people were of the ſame Celtic origin, ſpoke the Celtic language, and the firſt poſſeſſors of thoſe countries, appears by the Celtic names of places impreſſed on the very countries, and the following hiſtorial paſſages, viz. the Auſones in Italy, then called Auſonia, from Aulona in Peloponeſus in the days of Saturn, who were followed thither by the Oenotrians from Arcadia, who found there the Cimbrians or Chumbrians, Sabines and other aborigines or maritime migrators, deſcendents of the weſtern Celtes, Gauls and Titans; the Celtes or Phrygians, comprehending the Ionians, Trojans, Lydians, Myſians, Bithynians and Piſidians, with the Dioſcuri or Druids in the foreſts of Tarteſa or Spain; and the Gerions or Germans on the tracts of the Briges or Iberi of Spain, who were the Ever and Erimon of the Iriſh, and were drove out of Spain by Hercules Idæi, the Cretan, the Melichartus or the confines flood ſweeper, Midacritus or the Mediteranean Cretan, and ſon of Jupiter by Alcmena or the high water confine laces or havens; the colonies of Tyras, Briges, Allobriges, Hercules and Mercur Teutat or Gomer, Pluto, Mercury, and all thoſe colonies, ſuppoſed by Mr. Bochart to be Phœnicians, ſettled in different parts of Europe, except one on

the

the river Bætica in Spain. See Callimachus, Dion. Halicar. Polybius, Juftin, Cæfar, Diodorus, Tacitus, Berofus, Jofephus,
Tzetzes, Strabo, Solinus, Servius. Florus, Silius, Varro,
Stephanus, Elian, Ifidore, Oca Scarlenfes, Rhenanus, Fabri-
tius, the commentators on Jeremiah the fifth chap. Cluver and
other moderns. Hence Bribe, the water part fpring part;
Bride, the fide water part; Bridge, the fide water part edge;
Bridle, on the fide fpring water part; Brief, the fpring water
part part; Brier, on the water part; Bright and Brilliant, on
the water part fide; Brin and Brink, the water part rim, edge
or confines; Bring, the fpring part to the confines water part;
Brine, the inclofing water part; Brittle, on the water covering
fide; Brittle, on the water part fide, as an earthen veffel,
which is brittle and liable to cracks and fplits.

BROADWATER Suff. Ire. the furrounding water part fide
or country or broadwater. Broadwell and Broadftear Som.
Kent, the furrounding water part fide dwelling or country.
Brocard caftle Salop, the furrounding water part guard caftle.
Brochty-crag Scot. the furrounding water part or border above
the rock fide. Brokefby Leic. the border water fide habitation. Broket-hall Herts. the border water part fide hall.
Bromefgrave Worc. the furrounding border warden. Brom-
field on the Dee N-Wales, the furrounding water border
dwelling place or vale. Bromfiet Yorkf the furrounding water border fide place. Bromholme Suff. the furrounding water
place ifland. Bromley or Gerards Bromley Kent, Worc Staff.
the furrounding water part place or palace. Bronefcomb Corn.
the vale on the furrounding water part fide. Bronwen N-Wales,
a hill or country on the inclofing fpring. Brotherton Yorkf. a
town on the furrounding water part or border fide. Brough
and Brough in Yorkf. W-mor. Lanc. the furrounding water
part confines or border, and the border town. Brounfover
War. over the border river. Brovonacum, Brocovarr, Bro-
coniacum or Brougham W mor. the border fpring water part
vale or dwelling place. Hence Brooch, from the furrounding
water part or border; Broad, the furrounding water part fide
or country; Breech, the furrounding water part edge; Brock,
the furrounding water part confines, where the badger ufually
frequents; Broil, the furrounding water place food; Brood,
the furrounding part fide or brook, where fifhes breed; Brook,
the furrounding water part confines or fide water; Broom,
but the furrounding water part or con... Broth, the wa-
... the water part fide; Brothel, on the ... ding water
part fide or mal houfe as dwelling on ... ... from the
furrounding water, brothel and ... neighbouring ... ... a

2

of the furrounding water part; Brow, the furrounding water part fpring; Browfe, the furrounding water parts or country fmall fprings.

BRUCE, the fide fpring water part. Brudenel, a place on the fide fpring water part. Bruen abbey Oxon. the incloling fpring water part abbey. Bruges in the Low Countries, the lower confines fpring water part. Bruitton Som. the fide fpring water part town; it being the border, as well as the place of refidence of the Moions, the weftern feamen or Cornifh-men. Brumford N-hum. the furrounding or middle fpring water part ford or way. Brumham Wilts, a vale on the middle or divifion fpring water part. Brun Alben, Brunfwic, Brung-wen and Brunlefk Scot. Germany and N-Wales, the incloling water part hill, fide ftreet or fortification and moted or palace part. Hence Bruife, the fide fpring part; Bruit, along the fide fpring; Brumal, on the furrounding fpring part; Brunt, at the incloling fpring part; Brufh, the incloling fpring part fide; Brute, the animal fpring part.

BUALT or Bulleum Silurum S-Wales, the fpring water part fide, hill or dwelling place. Bubwith Yorkf. up the fide fpring part or dwelling place. Buchan or Buquhan Scot. on the incloling fpring water part. Buckenham Bucks. and Norf. a vale on the incloling fpring water part. Buckhurft Suff. the confines fpring water part foreft. Buellon Der. the fide fpring water part town. Buda in Panonia or Hungary on the Da-nube, the fpring water or dwelling part. Budley Devon, the fpring water fide place or dwelling place. Buers Suff. the fpring water part fide. Bueth Cumb. the fide fpring part. Bugden Hunt. the confines fpring fide, den or vale. Buley W-mor. the fpring water part place or habitation. Bulkly Chef. the fpring water place fhut place or dwelling place. Bullen or Boulogn in France, the fpring water place bank. Bulley Nott. the fpring water part place or dwelling place. Bullingdon Hants, a town on the confines fpring water place. Bulmer N-hum. the mere fpring water place Bulneis Cumb. the fpring water place on the promontory. Bulverhith Suff. the fpring water place paffage or ford. Bumfted Eff. the fur-rounding fpring water part ftation. Bunbury Chef. the incloling fpring part burrow. Bunraty on the Shannon Ire. on the incloling fpring part fide. Burcelter Oxon. a city on the fpring water part. Burford Oxon. and Salop, the ford or way on the fpring water part. Burgh, Byrig, Burrium, Bury or Vertura Hants, Surry, Suff. Der. Yorkf. W-mor. and Germany, on the fpring part confines or dwelling place. Burghate Yorkf. the fpring water part confines, fhut or dwelling place. Burgh-

ley,

ley, Burley or Burleigh Rut. N-hum. the fpring water part confines, fhut or dwelling place. Burgfted Eff. the inclofing fpring water part or burrow ftation. Burgundi France, on the inclofing fpring water parts confines, fides or burrows; who are fuppofed to be of the Vandal or vale race, but not the Gothic poffeffors of the fea coafts. Burghwafh Suff. the fpring water fide burrow or dwelling place. Burian or Berion Corn. on the fpring part or dwelling. Burn or Bourn Dorfet, Gloc. Linc. the inclofing fpring part neighbourhood or country. Burnel Salop, on the inclofing fpring water place or dwelling. Burnham Bucks, Norf. a vale or home on the inclofing fpring water part. Burnhop Dur. the fpringing up water part. Burnfal Yorkf. the inclofing fpring water fide place. Burnt Pelham Herts, the inclofing fpring water part fide fartheft vale or dwelling place. Burrium or Ufke S-Wales, Lanc. on the fpring water part. Burrowbridge, field, and ton Yorkf. Nott. N-ham. the fpring water part bridge, field, or town. Burwell Camb. the fpring water part dwelling place. Bufby Nott. Yorkf. the fpring fide part or habitation. Bufhbury War. the fide fpring part burrow. Bufygap N-hum. the fide fpring part gate or gap. Butterby Dur. the fide fpring water part habitation. Buttermere Scot. the fide fpring water part mere or lake. Buttington N-Wales, the inclofing fpring water part fide town. Buvinda or Boyne Ire. the edge or confines fpring water part. Buxton or Buccton Der. the confines fpring water part town. By, Bye, and Byan Norf. Linc. the part, fpring part or habitation. Byfleet Kent, the flood fide part. Hence Bub, Bubble and Bubby, from its fpringing up; Buck, the animal fpring part incloing; Bucket, at the fpring fide; Buckle, the buck place or a lock of hair; Bud, the fpring part divifions; Budge, the fpring part edge or action; Budget, at the fpring part fide; Buff, off the animal fpring part; Buffet, the animal fpring part at; Bug, the fpring confines part; Bugbear, the fpring confines part bear; Build, on the fpring part fide; Bulge, on the fpring water place edge; Bulk, the fpring water part fhut covering or incloure; Bull, the high animal fpring part; Bulwark, on the inclofing fpring water; Bum, the furrounding fpring part; Bump, a part on the furrounding fpring part; Bumper, a water part on the furrounding fpring part; Bunch, a fhut on the fpring part; Bundle, the inclofing fpring fide piece; Bung, the fpring part in the inclofing water; Bungle, the fpring part in the inclofing water place; Bun, the inclofing fpring part; Bunter, the inclofing fpring part fide water part; Buoy, the furrounding water fpring part; Bur, the fpring part; Burden, on the fpring part fide; Burgage, the burrow pledge; Burgbote, a boot or recompence for guarding the borders; Burn, the incloſing ſpring water part; Burſe, on the ſpring water part ſide; Burſt, below the ſpring water part ſide; Bury, on the ſpring water part; Buſh, the ſpring water part ſide; Buſhel, on the ſpring water part ſide; Buſs, the ſpring parts ſides; Buſtle, on the ſpring ſide part; Buſy, the ſpring ſide part; But, at the ſpring part or the ſide ſpring part; Butcher, the animal ſpring part incloſer; Butler, the man on the butt; Butment, the boundary ſide part; Butter, at the ſpring water part; Buttock, the ſpring part ſide covering water part; Button, on the ſide covering part; Buxom or Bucçom, the ſurrounding ſpring water part ſhut; Buy, the ſpring parts; By, the part or ſpring part.

CABALI or Vellai a town of Languedoc on the Rhoſne in Narbonne Gaul, the incloſing ſ,ring water place or vale. Cabala, a myſterious uſe of letters amongſt the Jews, ſomewhat reſembling the hierog yphics of Mercurius Triſmegiſtus and the reſt of the Caberi, is named from the letters p or b and c with their ſpringing letters which are expreſſive of and deriving their origin from the ſea and river diviſions of countries and the parts of generation, which flow like rivers into the ſea. Caberi or Cabiri, the incloſing ſpring water parts or the ſea coaſt coaſters or keepers who were Phrygians, the anceſtors of the Britons and the deſcendants of Gomer, and ſaid to be four in number, namely Ceres or Cybeli and Proſerpine her daughter by Jupiter, Pluto or Dis and Mercury or Caſmillus, who were probably as many colonies of Phrygians and Samothracians that came by ſea into Gaul and Spain. Hence Cabal, Cabbage, Cabin, Cabinet, and Cable, an incloſed or ſecret part or place; as Cackle is, from being cloſe or ſecret.

CADBERY or Cathbregion Som. the water port or Britiſh incloſures, borders or confines or the weſtern water part confines or garriſon; it being the Camalet, leet or palace of king Arthur and other Britiſh princes. Cader Scot. the incloſing or ſtation water part. Cadi of the Turks, the keeper or judge of a town or diſtrict. Cadleys N-Wales, Ire. at the incloſing water place ſide place palace or leet. Cadmus of Tyre, the ſurrounding water confines keeper or chieftan. Caduceus the rod or ſceptre of Mercury, a ſymbol of his being the defender and keeper of men and countries by meres, without going to war. Cadwellon of Wales, the incloſing ſpring water place bank guardian. Hence Cadence, from the fall of waters; Cadet, a centinel or keeper.

CAER England, Wales, Scot, Ire. an incloſing water port

or a city or part fhut by water, walls or political divifions, as fhires or çhaers. Caer-Caradock-daf or dif-gae-guidi or Inch-keith-laverock or Carbantoricum-Marthen-Narvon-Palladur-Voran, Moran or Glanaventa, Wales, Salop, Scot. Som. N-hum. the city of the border water or Severn, the fide flood part, inclofed part, the fide or inclofing fpring part, the con-fines or inclofing fpring water place, on the fea fide, on the fea, of Pallas or Minerva on the edge fpring water place or pa-lace, and the furrounding water part bank. Hence the names of the radical Car, a contraction of Caer.

CÆSAREA or Jerfey, a Britifh Ifland and towns of Cappa-docia and other places, a city or part on the fide water. Cæfaro-magus or Chefford Eff. France, a city, ftreet or way on the fide furrounding water, and not the city of Cæfar any more than Drufomagus is the city of Drufus, but the boundary water part gate. See for the names of this radical under Cas, and Cag, Cage and Cake fignify, the inclofing fides or fhuts or a thing inclofed or fhut on all fides.

CAISHO, Coffi, and Cantihobery Herts the inclofing or fide fpring water part. Caius Bericus, a Britifh chieftain, the inclo-fing water part or Britifh chief. Hence Cajole, Caiffon and Caiff, as furrounding, inclofing or receiving a perfon or a thing.

CALAIS or Portus Icius or Ifcius in Belgic Gaul, the lower or ifland confines water place or part. Calaterium Nemus Yorkf. the inclofing water place covert or Celtic grove or foreft, or fuch places as were ufually left wild and uncultivated on the borders of countries. Calcaria or Tadcafter Yorkf. a city on the inclofing water place. Caldecot on the Severn fide S. Wales, the inclofing water place fide coaft. Calder Yorkf. the inclo-fing water part. Caldey ifland in the Severn fea, the inclofing water place fide. Caldftream Scot. the inclofing water place fide ftream. Caldwell Yorkf. the inclofing water place fide fpring or dwelling place. Calebeg or Callebec Ire the inclo-fing water place brook. Caledon War. the inclofing water place town. Caledonia Scot. on the inclofing water places, fides, banks or hills, or hills and dales or clan country; the genuine Caledon Scots being the fair complexioned Gauls mentioned by Diodorus, who made many migrations and re-migrations to and from Denmark, and from whom the Kelleys, Shelleys, &c. Callan, Cilan and Calne Wilts, Wales, Ire. the inclo-fing water place bank or fide. Callendar Scot. on the inclo-fing water place bank fide. Calliope one of the mufes and mother of Orpheus, the inclofing water place circle part. Cal-fhot caftle Hants, the inclofing water place fide fhut or caftle.

Calthorp

Calthorp Suff. the inclofing water part gate. Hence Calafli, the lefler fhut; Calculate, at the inclofing fpring water place fide; Caldron, on the inclofing water place fide; Calender, and Calends, on the inclofing water place bank fide, Calf, the fhut or fide part; Calid, the fide inclofing water place; Call and Calling, on the inclofing water place or mouth; Callous, the inclofing fpring water place; Calm, about the inclofing water place; Calumny, calling on the furrounding water part; Calx, on the inclofing water confines, which was the ufual places for all forts of ftones before the rocks were broken and ftripped of their coats, and from whence all the names of ftones feem to be derived. See Cla.

CAM, Camb, Cambo, Cham and Cambodunum Corn. Gloc. Camb. Yorkf. Vindelicia, the furrounding water part, the crooked part, the ham. home or vale part. Camalet, Chamelot or Hamlet Som. Eff. Scot the furrounding water place fide, leet or palace; from their being the feat of Cunobelin, Arthur, and other guardians of the boundaries or chieftains, and in a fecondary fenfe, the place of war or the military town, as were Camulon in Scotland, Camuledun in Effex, and Chamalot in Somerfetfhire Cambeck Cumb. the furrounding water part brook. Camblan a river of Cornwal, where Arthur and Modred had an engagement, the furrounding water part bank. Camboritum, Cantebrige or Cambridge and Gloc. the furrounding or inclofing water part fide edge, bridge or ferry. Camden Gloc. the furrounding water fide, vale or town. Camelford, the furrounding water place ford or way. Camalen or ford Gam Helen N-Wales, the furrounding water part bank or hill way. Camerium a city of Umbria or Chumbria in Italy, on the furrounding water banks or hills. Camifham Herts, the furrounding water fide vale. Hence Camel, crooked or fhut upon; Camifade, inclofing the fide; Camlet, furrounding the fide; Camp, a part furrounding or inclofing; Campaign, in the camp. See Cham, Gam, Kam and Sham.

CAN W-mor. the inclofing water. Candale or Kendale, the inclofing water vale or dale. Cane or Canock wood Staff. the inclofing water confines wood. Canefield Lane. the inclofing water dwelling place Candida Cafa Scot. the inclofing water fide part of the lower confines. Canford Dorfet, the inclofing water ford or way. Cangani, Cangi or Langanum Som. Suff. Chef. Wales, the inclofing water part or bank, which like the other names of places being appellative, occafioned the doubts about Oftorius's march againft the Cangi. Canings, Canington and Canfham Som. the inclofing or border

der water vale, town or hundreds, the extensive woods and commons in these parts of Wilts, Canoe, Gedney moore, Cadbery, and Ringwood, being probably the boundary betwixt the western and BelgicBritons and the camalets and leet places being situated along the frontiers. Canonium or Chelmsford Ess. on the inclosing water ford or town. Cantaberi Spain, Kent, Camb Ire and Brigantes of Yorks.&c on the inclosing water part or foremost coasts, the first possessions or Phrygian possessions. Canterbury Kent, the Kentish or first burrow. Cantire Scot. the inclosing water or foremost possessions or country. Cantium or Kent, the inclosing water side or foremost land or coast. Cantre, Chantred, and Hundred, the side inclosing parts, water parts or possessions, or the possessions of 100 freeholders, householders or farmers, situated within a district of cantonment or commot of four sides towards the sea and rivers, which were politically subdivided into tythings of 10 freeholds or households each, with proper cells, convents, monasteries, priories, cloisters, leets, druids, bards, barons, lords, abbots, monks, priors, prebends, canons, and priests, for the purposes of religion and government, and whose names are expressive of their situations, as monasteries cells and monks on the sea side, priors and prebends at the spring head, and convents, cloisters and canons on the river sides, and so their stalls and seats in the convents or cathedrals seem to have been disposed, and thence the distribution of the people into families, tribes and clans, and perhaps of the Roman legions into cohorts, centuries, and decuries; the decury being the same as the tything, the hundred as the century, the cohort as a whole canton, and the legion as four cantons, or a province of an hundred townships each. From this division of countries our numbers or method of reckoning seems to derive their origin; as, Un, or One, the inclosing spring; Dau, or Two, the spring division or divided sides; Tri, or three, the boundary spring, its bank and bank side or possessions, which in the Welsh is tri or tir; Pedour, or Four, the surrounding water part or border of four sides; Pimpe, or Five, the dwelling or inclosed part, farm, fief of one possessor, or freeholder; or village; Six, the sea side or confines; Seven, the standing or resting part, or a haven; Eight, an island, or a part surrounded by water, or an octo; Nine, the rain or clouds; Ten, the high covering, the sky, and fair weather, whose models were the division or creation of the universe, Eden, and other parts of the globe, which numbers and letters seem to explain. Canvey or Convenos Island, Ess. the inclosing or confines spring, or confines water part. Hence our conjugating verb Can, as getting

within a

within the inclofing part; Canal, the inclofing water place, from its being the boundaries of countries; Cancel, its deftruction; Candid and Candidate, at the inclofing water or border fide or divifion; Cann, the inclofing or fhutting the water in. Canon, on the inclofing water; thence Cannon, a large gun; Canton, on the inclofing water fide divifion of one hundred towns, poffeffions, or townfhips. See Chan, Gan, Han, Kan, and Shan, the fpringing radicals of Can.

CAPEL, Wales, on the inclofing water place, or an inclofed place, or a chapel, which at firft were an inclofure without a covering. Capitania, Italy, on the fide, or below the inclofing part, cape, or head land. Hence Cap, the inclofing or covering part; Capable, the inclofing part able; Cape, the covering part; Caper, the covering water part; Capitulate, on the covering or border place fide; Captain or Capture, on the border part fide; Capuchin, the head covering; Chap, the inclofing water part or a chink; Chapel, on the inclofing water part; Chapman, the inclofing water part man or factor on the border water part, where goods, cattle, and wares were exchanged or fold; Chapter, on the inclofing water part fide. See the other fpringing radicals of C.

CARATACUS or Caradoc, the Silurian chief, on the fide water or Severn coaft chieftain, or plurium gentium imperator, as he fays in his fpeech before Clodius; he being fo as guardian of the Severn border, and commander in chief of other gentes or tribes, though natural lord perhaps only of his own clan. Carbantoricum or Laverock, Scot. the inclofing fpring water place or border vale city. Carbrey, Ire. the water part confines or city. Cardigan, S-Wales, the inclofing water fide city. Caren, or Carn, Corn. and Wales, on the inclofing water, or a rock or hill; as Carn-ladron, the hill of thieves; Carnedh, a heap of ftones; Karn ar oy ben, a heap of ftones on thy head. Carenton, Som. a town on the inclofing water, where Carentocus and Decombes are faid to have landed from South Wales. Carerbrooke, Hants, on the inclofing water fide brook. Carefdike, Linc. on the inclofing water fide dike. Carefwell, Staff. on the inclofing water fide fpring or dwelling place. Carew, S-Wales, on the inclofing fpring. Carlingford, Ire. the inclofing water place, ford, or way. Calile, Lugubalium or Luguvalium, Cumb. the inclofing water place ifle, or a vale on the inclofing fpring water place. Carr, Dorfet, on the inclofing water. Carram and Carrow, N-hum. on the furrounding border water or fpring. Carrict, Scot. the upper border fide. Carriden, Scot. N-Wales, the border vale. Carthage or Carthago, the inclofing water fide edge, confines, or city. Cartifmandua or Cortifmandua, the border below the fide of Mandua or Idumanum, a harbour or capital manfion of the Ifceni; fhe being chief of the Coritani. Carthmel, Lanc. on the furrounding water place or fea confines. Carvillus, a Kentifh chieftan, on the inclofing fpring water place warden or chief. Hence Car, Cart, Chariot, Carver, Cargo, and Carry, on the inclofing water; Cardinal, on the inclofing water fide; Care and Charge, on the inclofing water part confines or edge; Carnal, the upper inclofing place or frefh; Carol, on the inclofing water or country; Carve, the inclofing fpring part; Character, the inclofing or boundary water part fide; Charm, the furrounding water part; Chart, the inclofing water part fide, or a delineation of coafts; Charter, on the inclofing water fide; Carue, Caruage, Thong, Hilda, Hida, Soc and Soccage, the inclofing fpring, or furrounding water fide or edge; fuch was the portion of land that was alſotted by Vortigern for the defence of the borders to Hengift, which he inclofed with the flour and carue, thong and hide, fignify a boundary river.

CASHALTON, Surry, a town on the fide or lower inclofing water. Cafmillus, or Mercury, on the furrounding water or fea lower borders. Caffandrea, Suff. Greece, the lower inclofing water part confines. Caffel, Germany, the fide, lower or leffer water confines or city. Caffii, the ifland or Ifceni confines or inclofing water; Caffibeline, the Caffii or Ifceni water place bank keeper, warden, or warrior. Caffil, Ire. the lower inclofing water place. Caffiterides, the lower inclofing water part fides or iflands, or the weftern iflands, but no particular iflands; nor did the ancients mean any fuch, but fome iflands in the Atlantic, from whence the tin was brought, and according to Diodorus, who muft have known f om whence the tin was conveyed into Gaul, and thence by land to Marfeilles, where the Phoenicians had it, it muft have been fhipped on board by the Veneti at the Ifle of Wight, and it feems probable that the tin ore was at firft found along our weftern fea fhores, without digging for it. Cafter, Ccafter, Chefter, or Venta Ifcenorum, Norf the fide water part inclofure, fortification, caftle, or city of the Veneti or Britons of Gaul, in the country of the Ifceni or Saxons, the defcendants of the Keni or Kentifh, and from which commixture came the Brigantes of Yorkfhire, and other parts of England and Ireland, the Aborigines of England, according to Cæfar, and according to Pliny, of the fame origin as the Britons and Pontes of Gaul, and part of the Pontic or Wapentic province, was at firft within their limits in Britain; though the Logrian province of the Coritani afterwards extended itfelf pretty far into

their

their limits. Casterley, Wilts, the lower inclosing water side, place, or the castle place. Castor and Pollux, the lower confines border, and the upper surrounding light part or sky. Castle Ashley, Colwen-comb-corndocken-dinas-dun-steed-thorpe, N-ham. Wales, Wilts, Scot. Ire. Cumb. Bedf. an inclosure, fortification, or castle on the lower part, on the inclosing spring side, in the vale part, on the top of a hill, of the city, of the hill or town, of the side or station part, and water gate part. Castrum Alatum, Dun Eaden, and Edenburrow, Scot. on the vale side castle or city, the water vale hill or town, and the vale or lowland burrow. Hence Cascade, Case, Casement, Cask, Casiock, and Castle, as being places and things inclosed, and Castellan, and Castleward, the keeper of a border castle, and a compensation or payment assessed for watching and defending the borders and border castle. And many names springing, derivative, or inflecting from C, may be seen under Ch, G, H, K, Sh.

CATERICTONIUM, or Caterick, Yorkf. on the inclosing spring water, upper, or Richmond border, and from the great fall of water there in the river Swale, a cataract. Carterloch, Ire. the inclosed water place confines part. Catesby, N-ham. the lower inclosing water side part or habitation. Cathnefs, Scot. the inclosing water side promontory or lower part. Catlidge or Kirtling, Camb. the inclosing water side edge or border. Catmose, Rut. the inclosing water or border's surrounding side. Catterhal, Surry, on the lower inclosing water side. Catti and Cattieuchlani of Germany, Herts, Bucks, Bed. &c. the inclosing side, or spring water side banks, or the keepers of such, those of Germany and Britain being probably on the Watling-street, and such another street near the Rhine. These people and the Coritani seem to have migrated into North Wales, and from thence to Dumbritton in Scotland, and are probably the Stratclwyd Britons, Novantes, &c. as the names on Watling-street and other parts of the country seem to indicate. Cattihill, near Watling-street, Staff. the side inclosing, or Catti hill. Cattimanus, Germany, the mere or surrounding water side keeper. Cattivellan, the inclosing spring side keeper. Catus Decianus, the procurator, put to flight by Boadicea, the boundary water keeper. Hence Cat, the keeper, or house-keeper; Catacombs, the keeping shells or cells; Catalogue, keeping or containing the discourse; Catch, inclosing at the sides; Cater, on the inclosing water side; Cathedral, on the inclosing water part side; Cattle, at the inclosing water side or vale place; Catholic, as being general or common to all countries, or the whole world.

CAULI,

CAUCI, or Meath of Ireland, Frizeland, and Weſtphalia, the incloſing ſpring water confines or ſide. Caude, or Caude-bec, Cumb. the incloſing ſpring water part or brook. Cavel, or Gavel, Corn. Kent, Wales, an incloſed water place a boundary ſet out or a holding. Caverſfield, Bucks, the in-cloſing ſpring ſide, vale, or home. Caun, or Cavan, Som. Ire. Spain, Mauritania, and Cilicia, on the incloſing ſpring water ſide or incloſure. Caurſe caſtle, Salop, the incloſing ſpring water ſide caſtle. Cawood, Yorkſ. the incloſing ſpring water ſide or wood, or an incloſed or covert wood. Caxton, Camb. the incloſing water confines town. Cayhill, Wilts, the incloſing water high place or hill. Hence Cave, Cavity, and Cavern, a ſhut or incloſed part; Caveat, at the cave, ſhut, or ſtand; Caul, ſhut upon; Cauſey, the incloſing ſpring ſide; Caution, on the incloſing ſpring ſide.

CEADA, or Ceda, the incloſing water parts or ſhuts. Ced-walla, and Cadwallan, the incloſing ſpring water place banks, or their keepers, both in the Saxon and the other Britiſh dia-lects, which, as to the names of places and perſons, are funda-mentally the ſame; for thoſe who, on the removal of the Ro-man empire from this iſland, aſſumed its dominion, being moſtly the families in power under the Romans, joined by o-thers, who had been forced from hence to the continent and Ireland by the Romans, took to the names of the places they poſſeſſed, or thoſe of their offices, which, according to their definitions in this and a former eſſay, appear to be original Britiſh, Corniſh, Belgic, Logrian, Welſh, Northumbrian, Erſe, Lingua Romana, and other Britiſh names, and not that gramatical dialect compoſed and attempted to be introduced by Bede and other Saxon writers. Celtæ, Celd, Cell, Kelley, Shelley, Calle, Calleare, Salene, Calledonia, and Clana, are the various ways of expreſſing the incloſing water place ſide, confines, banks, cells, groves, covert place, and the habita-tions of druids, hermits, monks, and other recluſe perſons in the earlier ages, in Aſia Minor, Greece, Italy, Spain, Gaul, Britain, Ireland, and other parts. Celtiberia and Celtica, on the Iber in Spain, the ſpring water part, incloſing water place, or the Phrygian cells or harbours. Celurca, or Montroſs, Scot. the incloſing water place at the ſpring confines, cells or har-bours. Cenimagni, and Cenomania, of Suffolk and the reſt of the Iceni country, Lombardy, and Gallia Celtica, the ſur-rounding water incloſing or foremoſt part. Cenio and Cenio-nis Oſtium, Corn the incloſing water, or lower incloſing water mouth, haven, or gate. Cerberus, the three headed dog, ſaid to have been bound in chains, and thrown out of hell

by

by Hercules, the lower border water parts, or countries of the Titans, which Hercules drained, by drawing rivers through them. Cerdick, Cerdicford, and Sands, Hants, Norf. the inclosing water or sea side confines, way, sands, or the names which a Saxon chieftain assumed therefrom. Ceres or Isis, the lower water borders. Cernes, or Cerones islands of Scot. the lower inclosing water side or borders. Hence Ceafe, the inclosing water or sea side or haven; Ceil, an inclosure upon; Celebrate, at the cell water part; Celi, Celery, and Cellar, the cell or a low inclosed place, or at the water place side, which is low; Cement, the surrounding water on the sides, which joins together the parts of a country; Censor, on the inclosing water side man, who had the power of taxing among the Gauls, as well as the Romans; Centre, the side inclosed part; Centurion, on the inclosed side; Century, the inclosing side; Cerate, inclosing or shutting at the sides; Ceremony, on the surrounding water part or sea border; Certain, on the inclosing water side; Cessment, Cessation, Cession, and Session, on the inclosing water side or country.

CHAD, Chadshunt, or well, War. Salop, the inclosing water side, high spring, well or dwelling place. Chaldæa, of Assyria, the inclosing water place side, upper part or country. Chalgrave, Oxon. the inclosing water place warden. Chamber, Chef. the surrounding spring water part, hut, shut, cell, or other inclosure. Chanan, on the inclosing water end, or upper part. Charing cross, the inclosing water border cross, or the carrying cross, from its being the place of importation and sale of merchandises imported from the Thames. Charlcot, War. the inclosing water place confines, coast, or wood, Charleton, Worc. the inclosing water place town. Charmouth. Dorset, the inclosing water part mouth. Charn wood, Leic. the inclosing water or confines part wood. Charon, the ferryman of hell, on the inclosing or border water. Chastleton, Oxon, on the lower inclosing water place town or cattle. Chatham, Kent, the inclosing water side vace, home, or shut. Chatsworth, Der. by the inclosing water side lower side. Hence Chace and Chafe, the inclosing water side; Chaff, the inclosing or shell part; Chain, the inclosing water; Chalk, the inclosing water place confines, where the stones were first met with; Challenge, and Calling, the inclosing water place or borders; Chamber, the surrounding part or water part; Champain, the surrounding parts or country inhabited; Champion, on the surrounding water or borders, which were men of chivalry or knights; Chance, on the inclosing water side; Chancel, the inclosing cell; Chancellor, the inclosing cell

man;

man; Change, the incloſing water confines, where goods were exchanged and altered in their property; Channel, the incloſing water lowell or deepeſt place; Chaut, on the incloſing water ſide or chauntry; Chaos, incloſed on all ſides; Chap, and Chaps, the incloſing or divided part; Chapel, on the incloſing water place; Char, on the incloſing water or carrying; Charm, on the ſurrounding water part; Chaſt, the incloſing water ſide low; Chat, at the incloſing water part or mouth; Chaw, or Chew, the incloſing water part or mouth ſpring.

CHEBSEY, Staff. the incloſing water part ſide, lower part or Iſland. Checkley, Staff. a place on the incloſing water place. Chedeſley, Worc. the incloſing water place ſide lower place. Chelmer, Eſſ. the incloſing water place mere, it being the boundary of the Trinobantes. Chelmsford, its ford, way, or ſtreet. Chelſey, Midd. the incloſing water place ſide lower place or iſland. Chemleigh, the incloſing water ſurrounded place, as an iſland or headland. Chepſtow or Caſwent, South Wales, the lower incloſing ſpring ſide or ſtation on the Veneti lower confines. Cherdſley Bucks, a place below the border water ſide. Cheren or Cornwal, on the incloſing water place. Cherford, Devon, the border way or ford. Chertſey, Surry, the incloſing water ſide or iſland. Cherwell, Gloc. the incloſing ſpring water or dwelling place. Cheſham Bois, Bucks, the incloſing water ſide vale wood. Cheſter-field-ton in the ſtreet in the wall, Bedf. Der. N-ham. War. Dur. N-hum. the incloſing water ſide fortification, incloſed, or city dwelling place, town, in the ſtreet, and on the wall or ſpring water place. Cheſwerden, Chef. the city or incloſing ſpring water ſide vale, den, or foreſt. Chettelhampton, Devon, on the incloſing water ſide vale part town. Chettlingham or Chillingham, N-hum. the incloſing water place confines vale. Chevin and Cheviot, Staff. Yorkſ Wales, Ire. the high incloſing ſpring ſide, hills or ridges, which were ſuppoſed to have ſprung up like Marchey hill, Heref. Chevy Chaſe, N-hum. the incloſing ſpring water ſide, confines, or chaſe. Cheyneys, Bucks, on the incloſing water ſide. Hence Cheap, the incloſing water part, where things were cheap at firſt hand; Cheat, the incloſing water ſide; Check, the incloſing water confines; Checks, the incloſing water ſides; Cheer, on the cheeks or incloſing water ſides; Cherub, the upper borders; Cheſt, ſides incloſing the lower; Chevalier, on the incloſing ſpring water place man; Chevaux de friſe, incloſing the ſide parts.

CHICHESTER, Suſſ. the incloſing water or key city. Chickſand, Bedf. the incloſing water confines lower ſide. Chidley or Chudleigh, Devon, the incloﬁng ﬁde or joint place. Child-wit, the incloﬁng water place ﬁde ﬁp ing. Chi ham, Kent, the incloﬁng water place vale. Chiltern the ﬁde or incloﬁng high land; the Chi tern hills being probably the boundary be-twixt the Cattieuchlari and the Boduni. Chin ligh, Devon, the furrounding water place edge or confines. Chipches, N-hum. the incloﬁng water part key or lower race. Chipen-ham, Wilts, the vale or dwelling place on the incloﬁng water part. Chipen Sodbury, Gloc. an incloﬁng water part on the furrounding ﬁpring water part. Chirbury, Salop, the border ﬁpring water or dwelling part. Chilk, N-Wales, the border water gate or ﬁhut. Hence Chicken, on the incloﬁng water ﬁhut; Child, the incloﬁng water ﬂowing diviﬁon, or upper race; Chill, on the incloﬁng water place; Chine, the fur-rounding borders or country; Chimin and Chiminage, the incloﬁng water edge or way, and toll paid for a way through the king's foreﬁt lying along the boundaries, which belonged to no private perﬁon, but were left as commons on the limits of countries; Chine, incloﬁng or ﬁhutting in; Chink, inclo-ﬁng in the confines; Chip, the incloﬁed water parts, or the leﬁer diviﬁons of countries into farms; Chirographer, on the border or country writer; Chiﬁel, the leﬁer incloﬁng water place; Chit, the firﬁt part or diviﬁon; Chivalry and Soccage, to which all the ancient Engliﬁh tenures have been reduced, ﬁubject to different ﬁervices due to the crown; on the incloﬁing ﬁpring water place, and the furrounding water part edge or con-fines, or its engagement, action or ﬁervice for the defence of the country, by and againﬁt the Saxons and others.

Cholmondley, Chef. the furrounding water part ﬁde, incloﬁng water place or vale. Chopwell, Dur. the incloﬁng water part ﬁpring or dwelling place. Chorley, Lanc. the in-cloﬁng water border place. Hence Choice, the circle ﬁde; Choke, the furrounding water ﬁhut; Choler, on the furround-ing water place; Chop, the incloﬁng water ﬁpring part; Cho-ral, on the border water; Chord, the border water ﬁde; Cho-rographer, an engraver or writer on the borders of coun-ties.

Chur, Som. the incloﬁng ﬁpring. Churn, Wilts, the in-cloﬁng border water. Chute, Hants, the incloﬁng ﬁpring ﬁde. Hence Chub, an incloﬁed or ﬁhut head or upper part; Chuffy, a ﬁhut part; Chum, jointly furrounded; Church, the incloﬁ-ing ﬁpring water confines or meeting; Churl, on the incloﬁng ﬁpring water place, or an earle; Churn, the incloﬁng ﬁpring water place; Chyle, the firﬁt ﬂowing or ﬁpring water.

Chygarth, Corn. the incloﬁng water place ﬁde incloﬁures.

Cilurnum, N-hum. on the inclofing fpring water place fide. Cimen, Suff. the furrounding water or fea edge. Cirencefter, Durocornovium, or Corinium, Gloc. a city on the inclofing water or border. Cisfbury, Suff. the leffer or fide inclofing fpring part. Hence Cider, the inclofing chief or joint fpring water part; Cincture, the inclofing fpring water part; Cingle, the inclofing water place; Cion, the firft in; Circle, the inclofed water place border; Circuit, the inclofing fpring border fide; Circumambient, about the furrounding fpring border part on the fide; City, the inclofing water or joint fide or poffeffion; Civil, a joint dwelling.

CLACKMANAN, Scot. on the inclofing water place confines part, haven, or harbour. Clacton, Eff. the inclofing water place confines town. Clan-car-don-Gibon, Ire. Surry, the inclofing water place fide or bank city poffeffions, or clan town, or weftern boundary. Clapham, Surry, the inclofing water place part, vale or home. Clare, Suff. Ire. on the inclofing water place. Clauth, Wales, the inclofing fpring water place fide bank, fence or inclofure. Claufentum, Trifanton, or Southampton, Hants, on the lower inclofing fpring water fide, a town on the lower fide water part, and the fpring fide or fouthern vale town. Claxton, Dur. the inclofing water place confines town. Clay, Norf. Salop, the inclofing or fhut water place. Claybrooke, Leic. the inclofing water place brook. Claycefter, Leogora or Leicefter, the inclofing water place city or borders. This was the metropolis of the kingdom of Leogora, originally comprehending the limits of Leicefter, Lincoln, Nottingham, Derby, and Stanford, and founded by the German Fife Burgenfes, and to which the limits of York and Chefter were afterwards added by the Seven Burgenfes, a mixture of the German, Danifh, and Roman, and they held a parliament of their own, without any connection with London, which in thofe days was a free town. Hence Clack, the inclofing water place confines or action, or the mouth action; Clad, inclofing the fide; Claim, the inclofing water place border or rim; Clamm, about the inclofing water place; Clan, on the inclofing water place; Clank, the mouth action; Clap and Clapper, the inclofing water place part; Clarencieux, the chief one of the inclofing water place; Clafh, below or at the fide of the inclofing water place; Clafp, the inclofing water place fide part; Clafs, below the inclofing water place fide; Clatter, at the mouth or border part; Claufe, the inclofing fpring fide; Clafure, the inclofing fpring fide fpring, which confined and divided the country into farms or leffer portions;

Claw,

Claw, an animal fhut; Clay, the water inclofing or covering part.

CLENT, Staff. the inclofing water fide or bank. Clethen, S-Wales, the fide divifion or inclofing fpring water place. Hence Clean, Cleanfe, and Clear, on the inclofing water place; Cleave, the inclofing water place part; Clement, the inclofing water place fide part. Clench, the inclofing water fhut; Clergy and Clerk, on the inclofing water confines; Clear, the inclofing on the fpring; Clew, the inclofing fpring.

CLIF and Clifton, Kent, Worc. Nott. the inclofing water place part and town. Clinton or Glinton, Linc. Nott. War. Staff. Wales, Ire. a town on the inclofing water place or the vale town. Clift and Clipfby, Devon, the inclofing water lower fide or part. Clithere, Lanc. on the inclofing water fide. Cliveland, Yorkf. the inclofing fpring water place or bank fide. Hence Click, the inclofing water place or mouth action; Client, on the inclofing water place fide; Cliff, the inclofing water place high part; Climate, inclofing the furrounding fides; Climb, the inclofing water place bank part; Clinch, inclofing a fhut; Clip, the inclofing water place part.

CLOGHER, Ire. on the circle inclofing or inclofed water place or lake confines. Clomel or Clonmel, Ire. the middle inclofing water place or bank. Clon or Clan-fert-lolan, Ire. on the inclofing water place fide part or bank. Clovelly, Devon, the inclofing water place, fpring or dwelling place. Cloveshoo, Kent, the inclofing high fpring water place fide. Clowcrofs, Camb. N-ham. Linc. the inclofing fpring water place crofling, crofs or paffage. Hence Cloak and Cloth, covering the fides; Clod and Clot, inclofed round or on all fides; Clog, inclofed or fhut fides; Cloifter, the inclofing water place fide part; Clofe, the inclofing fpring water place or fhut fides; Cloud, the furrounding fpring water inclofing or fhutting the fides; Clover, the inclofed or fhut fprings; Clout, a man fide inclofing or covering; Clown, a man on the inclofing water place; Cloy, the inclofed or fhut water place.

CLUN, Colun or Colnwy, Salop, on the inclofed fpring water place or vale. Clynog, N-Wales, the fea fide vale. Clytheren, Scot. Wales, the high inclofing water part. Hence Club, the inclofing water place fpring part, which join and blub; Cluck, the inclofing fpring water place or throat action; Clumfey, the inclofing fpring water place about the fides; Clufter, the inclofing fpring water place fide part.

COMBE, Devon, the covering fpring water part. Coberley, Gloc.

Gloc. the covering spring water part place. Cobham, Surry, the covering spring water part vale or dwelling place. Hence Coach, a covering sides or shut; Coal, on the inclosing spring water place; Coarse, on the covering water or sea side; Coat, covering the sides; Coax, the covering water confines side; Cobble, the covering spring water part place; Cob or Spider, the covering or web spring part.

COCCIUM, Cocar, Cock, Cockley and Cocker, Lanc. Yorkf. Camb. the spring water on or together on the surrounding water confines. Cockerfand, Lanc. the cocker fand. Hence Cock, the covering spring or animal; Cockle, the covering water spring place; Cockney, on the covering spring water.

CODENOR, Der. the covering water side on the border, or a wood on the border. Coderidge, Worc. the covering spring water part edge, ford or bridge. Codington, Chef. a town on the inclosing side or shut. Hence Cod, the covering water or sea part or division, or the side water covering; Coerce, lowering the water coverings; Coexist, together at the side water confines; Coffer and Coffin, the covering or shut part.

COGIDUNUS, chieftan of the Regni, the covering water confines man. Hence Cog, a covering action; Cogent and Cognition, a covering action on the side; Cohabit, at the covering water place spring part side or dwelling; Cohere, on the inclosing water; Cohort, on the surrounding water border side.

COITFALA, Corn. the covering water side, spring water place or vale. Hence Coif, the covering part; Coin, on the inclofure or covering; Coit, a covering or flat side; Coition, on the covering fide.

COL, Cole or Colle, Herts, Bucks, Scot. the inclosing water place. Colebrooke, Bucks, the inclofing water place brook. Colecefter, Colechefter and Colonia, Eff. N-hum. Germany, Italy, on the inclosing spring water place, bank or city. Colf-hull. War. Wales, the inclofing spring water place hill. Cobham, Midd. the inclofing spring water place vale. Collingham, Nott. the long inclofing spring water place vale. Collerford and Collerton, N-hum. the inclosing water place way or ford and town. Collivefton, N-ham. the inclofing water place fide town. Colrain or Londonderry, Ire. a country on the inclofing spring water place, and the spring water part, bank or lane town. Columb and Columben, Corn. Devon, the inclofing spring water surrounding parts vale or town. Colwal, S-Wales, the vale, or Wales inclofing spring water place. Colyhurft, Lanc. the inclofing spring water place fu-reſt. Hence Colation, on the incloſing water place ſide; Cold, the incloſing or ſhut water place ſide or diviſion; Cole, a ſhut place or part; Collar, incloſing upon; Collateral, on the incloſing ſpring water place ſides; Colleague, on the incloſing ſpring water place confines; Collect, at the ſpring water place confines, College, the incloſing ſpring water place edge; Colon, the incloſing water place within; Colony, on the incloſing ſpring water place; Colt, the incloſing ſpring water place ſide or diviſion; Column, in the ſurrounding incloſure.

COMBE, Cum or Comata, Devon, Wales, and France, the ſpring incloſing parts, vales, valleys, and other places of habitation from War. to Corn. Comius of Arras, the count of Arras. Combe Martin, Devon, a vale on the ſea ſide. Combe Neville, Surry, a vale on the ſpring water place, or a mere vale dwelling place. Combretonium, or Bretenham, Eſſ. the water part ſide vale. Compton, Oxon. War. the vale part town. Hence Comb, the ſurrounding part and vale, which was divided into ſmall diſtricts, like the teeth of a comb; Combat, the ſpring part at the ſurrounding water; Combine, the ſpring part in the ſurrounding water part; Come, to coomb or be at home or in the commot; Comfort, the ſurrounding water part way or ford; Comity, the ſurrounding water part ſide; Command, on the ſurrounding water part ſide; Commence, on the ſurrounding water part lower ſide; Commarchio, the ſurrounding water part mere confines, market or marking; Commerce, the ſurrounding water part mere ſide, the uſual market place; Committee, the ſurrounding water ſide diviſion; Common, the part on the ſurrounding water parts, which were left uncultivated; Companion, one on the ſurrounding water part; Compare, the ſurrounding water parts, whoſe ſides are alike; Compound, the ſurrounding water part pound; Comptroll, the ſurrounding water part ſide roll, court roll, roll rick, or the kingdom's roll.

CONCANI, Cangani, Cenaught, Cangi and Kendale, Suff. W-mor. Ire. on the incloſing water part confines or vale. Condate, or Congieton, Chef. the incloſing water part ſide or town. Condercum, Dur. the incloſing ſpring water part vale. Condover, Salop, the incloſing water ſide ſpring. Coner, Ire. on the incloſing water. Congerſbury, Som. the incloſing water border ſpring or dwelling part or burrow. Congham, Norf. the incloſing water confines vale. Conilah, Ire. the incloſing water place. Coningſton and Coniſide, Lanc. a town or ſide on the incloſing water. Conniſburrow, Yorkſ. the incloſing water ſide burrow. Channel, Scot. the incloſing water

ter place. Conwy, Conovius, Convenos, and Convey, Eff. N-Wales, the inclofing water fpring or dwelling place. Hence Con, on the inclofing water; Concatenate, the inclofing water fhuts below the fide; Conceal, the inclofing water fhut place or lower place, Concede; the inclofing water part lower fide, or to yield; Concern, on the inclofing water border; Conch or Shell, the inclofing water fhut; Conciliate, at the inclofing water place confines; Concord, the inclofing water border fide; Concur, the inclofing water current; Conduct, at the inclofing fpring part confines; Confederate, inclofing the water part and the fpring water part; Conflux, the inclofing fpring water parts flowing together; Confent, the inclofing water on the lower fide; Confign, the inclofing water fign or mark; Conful, on the inclofing fpring fide; Contain, the coaft within the inclofing water; Contend, the inclofing water fide end, where the river contends with the fea; Continent, the infide of the inclofing water; Contour, the inclofing water border or outline; Contract, the inclofing fpring water fide roll, or regifter of the feveral poffeffions or portions of land within the country; Controvert, turning the inclofing fpring water border; Convey, the inclofing fpring water confines or way; Conftable, the inclofing fpring water confines ftation place.

COPELAND, Cumb. the inclofing water part bank fide or land. Coptal, Eff. on the inclofing water part fide or coaft. Copti of Egypt, the inclofing water part fide. Coquet, N-hum. the inclofing water at the fpring confines. Hence Coop, fhut up; Coot, at the inclofing water; Cope, a covering part; Copious, the covering fpring part; Copfe, the leffer inclofing or covered part or covert; Copulate, the fpring part at the inclofing water place; Copy, the inclofing water part, which is a copy of the human parts.

CORBY, Cumb the inclofing water or border part or habitation. Coria Danmoniorum, Corn. Scot. the borders of the Danmonii, or fea fide inhabitants. Corberley, Gloc. the border fpring water part place. Corbet, Salop, the inclofing or border fpring water part. Corda, Scot. the inclofing water or border part. Corfe-Caftle, Dorfet, the border water part caftle. Coriondi, Wodix, or Corke, Ire. on the inclofing fpring waters, or borders, fides or poffeffions. Coritani, of Leic. Der. Nott Linc. and parts of Rut. Staff. N-ham. the inclofing fpring water fides, coafts, or borders, or the fpringing borders, as having quitted their ancient clan borders to extend their borders along the river fides or in vales. Thefe are genuine anceftors of the prefent Englifh. Cornavii or Cornabii

nabii of War. Worc. Staff. Salop, and Chef. the inclofing fpring water dwelling parts or borders. Cornhill, and Corndon, Chef. N-Wales, a hill on the inclofing water or border. Cornwall, Cornovaille, Cornubia, or Danmonium, the part or dwelling place on the inclofing or furrounding water or borders. Corfica, the inclofing water lower confines ifland. Cortiani, Yorkf. on the inclofing water fide or border. Corve, Salop, the inclofing fpring water or border part. Corvefham, and Corvefdale, the border water vale. Cory Malet, Som. the furrounding water place fide borders. Hence Corai, on the water place border; Cord, the inclofing part; Core, the inclofing fpring water; Corke, the firft inclofure or city, or on the water border; Coribantes, the priefts of Cybele, the chiefs of the bottom or vale borders; Corn, on the inclofures; Cornage, on the border edge, which the tenants were obliged to defend, but not by blowing of a horn, for Horn is from Chorn in Cornage; Cornucopia, the inclofing fpring water confines part, which refembled a horn; Corps, the inclofing water part fide; Corrode, the border water fide; Corrupt, the inclofing water fpringing part or the fide; Corfair, on the fide inclofing water.

COSHAM, Wilts, the furrounding water fide vale. Coffini, or Corini, Corn. on the lower inclofing water fide. Cofton, Som. the inclofing water fide town, or on the coaft. Hence Cofmographer, the furrounding or fea inclofing fide writing; Coft, at the inclofing water fide.

COTES, Leic. the inclofing water fides, coafts, or poffeffions. Cotgrave, Chef. the inclofing water fide warden. Cotenham, Camb. Yorkf. a vale on the inclofing water fide. Cotfwold, Gloc. the inclofing water fide wood or a fide. Cotus, the inclofing water lower fide chieftain. Hence Cot and Cottage, the furrounding water fide edge, or an inclofure covered on all fides in a fecondary fenfe.

COVENTRY, War. a town on the inclofing fpring water confines, or of the Veneti or Wenta, and the confecrating fpring town. Coverham, Yorks. the inclofing fpring water fide. Covert, Suff. the inclofing fpring water fide. Cowbridge, War. the inclofing fpring water confines town. County Palatine of Chef. Lanc. Dur. the remains of the ancient caldrons, or provinces inclofed by water boundaries, which the Romans made official for the defence of the kingdom againft the Irifh, Welfh, and Scots. Cowbridge or Bovium, Wales, the furrounding fpring water part, tranflated Cowbridge, or Cowy or Cowy, Midd. Surry. Suff. the inclofing fpring water fide. Cowdry, Suff. the inclofing fpring water part, town confi-

feffions. Cowel, Scot. the inclofing fpring water place. Cowes, Ifle of Wight, the lower water fide. Cowholm, Norf. the inclofing water place ifland, or furrounded dwelling place. Cowick, Yorkf. the inclofing fpring water ftreet, fhut or fort. Cowling, Kent, the inclofing water place bank. Cowlidge, Camb. the inclofing water place edge, it being on the limits of the Eaft Angles. Cowlock, Ire. the inclofing fpring water place, fhut or lock. Cowney, Camb. the inclofing fpring water confines. Coyne, Ire. on the inclofing fpring water. Coytnefs, Der. the furrounding fpring water fides. Hence Couch, the inclofing fpring water confines; Cove and Cover, the fpring water covering or confines; Covenant, on the inclofing fpring water fide, Covet, at the inclofing fpring; Covey, a covering; Cough, the inclofing water fpringing up; Could or Should, at the furrounding fpring water place; Councel, the inclofing fpring water place cell; County, Country, or Comty, the inclofing fpring water place fide or canton; Couple, the inclofing fpring water place or fides; Courage, the inclofing fpring water edge or action; Courfe, the inclofing fpring water to the lower fide; Court, the inclofing fpring water fide, where the courts leet and baron were held, and the rolls and regifters of the poffeffions kept; Coufin, on the inclofing fpring lower fide; Cow, the inclofing water fpring or animal; Coward, the cow warden, or one without a fpring; Cowflip, the inclofing fpring water place fide part; Coxcomb, the inclofing water confines vale part; Coy, covering.

CRÆCCA, Creach, Creake, Creke, Creioh, Creeanford, or Thorpe Greklade, Wilts, Kent, N-ham Norf. Der. Yorkf. W-mor. the inclofing water borders, fhuts or crekes, or the German or Legrian confines, countries, ways, ftreets, fords, and water gates; and at Creeanford Ethelwold raffed the Thames into Bredan, or Bretan, and, being ifland Danes, deftroyed it, according to Henry of Huntingdon.

CRAMOND, Scot. the furrounding water borders. Cranburn or brooke, Kent, Dorfet, Glocefter, the fpring part or brooke inclofing borders. Cranon, Yorkf. the inclofing fpring borders. Crawdenfhire, with, Weftm. the borders below the fpring water fide; Crawley, being the inclofing or border fpring water way or ford. Crewkerne, Somerf. the inclofing water fide confines. Crayke, York, the fpring water. Creyford, Kent, the inclofing water way or ford. Hence Crick, on the inclofing water, or the border water confines, Cricket, the inclofing ftreet; Crag, the inclofing water borders; Crick, the border, or where crams their, the inclofing or border county, in

in the inclosing water side; Crape, the border water part; Crash, the border water side; Crave, the spring border; Craw, the animal inclosure shut or bag, Crawl, on the Craw; Creak, and Creeke, the inclosing water border; Cream, about the inclosing water; Create, dividing the borders; Credit, at the border side; Crest, covering the lower side; Crib, the inclosing part; Crier, on the inclosing water border, where goods were sold; Crime, inclosing the high surrounding border; Crinkle, the inclosing water confines place; Crisp, the upper border side part; Criterion, on the inclosing water part or border mark; Critic, the border side action.

CROC, Croke, Crockhern and Crockern Tor, Chef. W-mor. Som. the water border, and the water border high part or tower. Crococalana, Ancaster, or Collingham, Linc. the inclosing border water bank, the inclosing spring castle, and the inclosing water confines vale. Croit castle, Heref. the border part side castle. Croeun, Linc. on the border spring. Cromarty, Scot. the sea side borders. Cromer, Linc. Scot. the surrounding water or sea border. Crosby, Lanc. Cumb. the border water side part or habitation. Crouch bay, Eff. the upper border water bay. Crowherst, Suff. the forest border. Crowther and Crowland, Linc. Yorks. the border water bank side or land. Croxden, Staff. the border water confines vale or forest. Croyden, Surry, the border den vale or forest. Cruicston, Scot. the upper border side town. Hence Croak and Crock, the water border; Crocodile, on the border water side place; Croft, the border part side; Crony, on the border one; Crook, the sea or water border; Crop, the border part; Crosier, on the border side; Cross, below the border side; Crow, the border spring or animal; Crowd, the spring water border side; Crown, on the spring water border; Crude, the spring water confines side; Cruise, below the border; Crump, the surrounding border part; Crush, the lower side or border; Crust, the lower side part; Cry, the flowing water.

CUCKAMSLEY, Berks, the surrounding spring confluence water place. Cuckfield, Suff. the inclosing spring confluence water dwelling place. Cuckmer haven, Suff. the inclosing spring water meeting the sea haven. Cuckney manor, Nott. the spring water confluence manor. Cudington and Cudlington, Surry and Oxon. a town on the inclosing spring water side. Culchit or Cilketh, and Culfurth, Lanc. Suff. the inclosing spring water confluence side, ways or fords. Cum and Cumberland, the surrounding spring water parts, bank sides, or possessions, or the possessions of the Cumberi. Cumbermere, Chef the surrounding spring water part mere or lake,

or the Cumberi meres; Cumbernland or Cumbernald, Scot·
the furrounding fpring water part, high land or ground·
Cumerford Wilts, the furrounding fpring water part, or Cu-
meri ford or way. Cunethio or Marlbro, Wilts, the inclofing
fpring water fide, or the mere water place border. Cuni-
glaffus, the inclofing fpring water place fide or chieftain, and
though Glafs in a remote fenfe fignifies green, yet no names of
perfons derive their origin from colours, but the names of
perfons and colours come from the names of places. Cun-
ningham and Cunington, Scot. Hunt. the inclofing fpring
water confines vale and town. Cunobeline, chieftain of the
Trinobantes, chief of the inclofing fpring water place or
Thames. Cungetorix, the inclofing fpring water border fo-
vereign. Cupre or Cupers, the inclofing fpring water part
fide. Curia or Curra Ottadinorum, N-hum. the borders of
the Tyne coafts, or the poffeffions of the Cumberi. Curetes,
or the Idæi Dactyli of Phrygia and Corybantes, on the lower
borders, the high parts on the water parts fides, and the lower
or bottom borders, or their guardians. Curraghmore, Ire.
the fea borders. Curlew mountains, Ire. the border fpring
water place mountains. Curwen or Culwent, Cumb. on the
fpring water place border. Hence Cub, the inclofing or firft fpring
of animal part; Cube, equal parts; Cuckold, a joint holding;
Cud, the fide inclofing fpring; Cuddle, on the fide inclofing
fpring; Cue, the inclofing fpring; Cull, on the inclofing
fpring water place; Cumber, the furrounding fpring water
part; Cumulate, extending the common or boundaries; Cun-
ning, on the inclofing fpring confines; Cup, the inclofing
fpring water part; Cupid, the inclofing fpring part fide;
Cur, a running dog; Curate, the inclofing fpring fide; Curb,
from running; Curd, without running; Curl, the inclofing
fpring water place; Curfe, below or the lower border. Curt, the
border fide; Cuftom, the inclofing fpring fide dominion; Cut,
the inclofing fpring divifion; Cycle, on the round of fpring
or high light.

CYMBRI, Chumbri or Umbri, Gomeri, Gallia Comata,
and Celtes, fignify the Gaulifh or vale Britains, the defcendants of Gomer and his three fons, Afkenas, Ripaath, and
Togarmah, whofe fift fettlements were within the father's limits in Cappadocia, Galatia, Paphlagonia, Pontus, Bithynia,
Myfia, Phrygia, and Troas, on the Euxine fea coafts. And
being poffeffed of the Phrygian letters, and the ufe of fhipping,
or the wings of Mercury, they were from thence conducted by
the Caberi, of whom Gomer or Mercury feems to have been
one, into the moft weftern parts of Europe, along the Adriatic

I

and

and Mediteranean coaft; planting, draining, cultivating, dividing, and diftributing the countries along the fea coafts, as they proceeded by the directions of the Caberi, and the form of the Caballa or Phrygian letters. Hiftorians, it is true, without confidering that there could have been at moft but five nations concerned in the plantation of Europe, namely Magog, Mefech, and Tubal, which the fcripture places as one nation in Ruffia and other northern countries, Maadai in Media, Tyras in Thrace, Javan in Greece and Afia Minor, and Gomer and his fons of courfe in the other parts of Europe, and that the foundation for any diftinction of nations betwixt thofe people muft have chiefly arifen from the different nature of countries, climates, employments, cuftoms and manners of the people, have difagreed as to the origin of thefe, as well as moft other European nations, by reafon of their meeting with fome of every fort of the Japhetan race almoft in every country of Europe, though Gomer and his fons were certainly the firft planters of the fouthern and weftern parts. But it being very immaterial from which of the fons of Japhet we derive our origin, or whether we made our firft migrations from Afia weftward with the fun, or through the northern countries, if, upon the whole, we appear to be of the Japhetan or Gentile race, deftined by providence to the poffeffion of Europe, or the ifles of the Gentiles, we need only obferve in general from hiftorians, that the Cumbri-Galli-Celtes appear to be originally but one nation, under the government of the Caberi or Druids, which extended over a confiderable part of Italy, and all Spain, Gaul, Germany, and Denmark, with the Britifh and other European iflands, and that from them, except as to Javan and the Turkifh ufurpation, thofe nations which have been fince formed within their limits are derivable. It may not however be improper here to obferve, as to the feveral routes from Afia into thefe parts, that they feem to have been prefixed according to the order in which Gomer and his fons were at firft feated along the Euxine fea, by the divifion or diftribution of providence; as for inftance, Togarmah, who was placed next to Magog in Afia, being the laft that quitted that country, and of courfe, the laft poffeffor in every country in Europe, muft confequently have been feated next Magog in Germany, according to the regular mode of migration and diftribution of countries. For the order and courfe of migrations and plantations of Europe, by Gomer and his bands, was firft by Gomer himfelf, with his own immediate tribes, down the Mediteranean fea into Italy, Spain, Gaul, Germany, and Britain; fettling colonies as he went in each country, and fetting out poffeffions for his fons

Atkenas

Aſkenas and Togarmah, who were to follow him. On Gomer's quitting Aſia, his ſon Aſkenas ſucceeded him in his poſſeſſions there; and after a length of time he left them to Riphath, and removed to thoſe which had been alloted him by his father in Europe. Riphath being by thoſe removals in poſſeſſion of the frontier countries, made a gradual migration from thence into the vacant poſſeſſions along the Adriatic and the Danube, leaving his Aſiatic ſtation wholly to Togarmah, who, after ſome commixture with the race of Javan, and founding a part of the Greek nations, at laſt purſued his father's courſe of migration, and ſettled colonies in Magna Grecia and other countries which his father had allotted to him in Europe. When any people became ſo enlarged as to want room, they ſwarmed forwards in the ſame regular order into the vacant poſſeſſions ſet out for them in the next country by their father, who was the firſt poſſeſſor thereof by the gift of providence; as for inſtance, Gomer with his followers poſſeſſed in Gaul the countries between the Loire and Rhine; leaving Vaſcones or Gaſcoigne and part of Spain for Aſkenas, and beyond the Rhine, oppoſite to the country of the Coritani of Britain, to his ſon Togarmah: and in Britain Gomer ſeems to have poſſeſſed Dover, and to have extended his limits on the ſea coaſts from the country of the Coritani to the borders of the Durotriges, and weſtward on Watling Street or the Iter Cymborum into North Wales, and from thence on the ſame ſtreet running along the weſtern ſide of the iſland to Dunbarton in Scotland. Togarmah with his Greeks took poſſeſſion of the country of the Coritani, and Aſkenas of the weſtern parts of the iſland, which from its ſituation was firſt diſtinguiſhed by the name of Britain. A ſort of coalition having happened between the Britons or Aſkenites, the lower confines men and the Gomerites, they promiſcuouſly ſent colonies into Yorkſhire and Durham, which extended on the eaſtern coaſts as far as the Firth of Forth, and from their being a commixture of Kentiſh and Britiſh came to be called Brigantes; from their ſituation on the Tyne, Ottadini, and Meatæ; from their dwelling in cells, vales, and lowlands on the ſea ſide, Selgovæ; and from their ſituation on the incloſing borders Gadeni. Northward of this Firth Togarmah took poſſeſſion of and extended it to Aberdeen, from thence to the Orkneys were colonized by the Gadeni and the reſt of the Brigantes. In the mean while Togarmah with his Greeks were buſily employed in extending their Coritani borders on the weſt, north, and ſouth, inſomuch as nearly to divide the iſland in the midſt, ſo as to form three kingdoms of it, correſponding with the ſtory of Brutus dividing the kingdom between his three ſons Locrine,

Alaanact

Albanact, and Camber, and calling them Logria, Albania, and Cambria, after their names, and marrying Locrine to Guendolena or the Veneti of Watling street, whereby Togarmah or Locrine became possessed of those countries; leaving the rest to his brothers Belinus and Brennus, or the Britons and Belgæ. But this marriage was probably somewhat of the same sort as that in Gaul betwixt Gomer or Hercules, the upper house, and the daughter of Alesia, the lower and the cleansing of the Augean stable, and driving Gerion or Torgarmah with his Greeks out of Spain, into those parts which had been allotted him in Germany, opposite to those of the Coritani in Britain. And this coalition of Togarmah and the Greeks seems to have been the chief foundation of our divisions in ancient Celtica, and the destruction of the Trojans and other Gaulish Gomerian nations, the elder and governing houses of Japhet, by the Greeks and Romans, assisted by the Togarmans and his Greeks, and becoming masters of that part of this island, which the Welch call Loger, the German vale confines, where many of those Greeks and Roman families, as well as Togarmans remain to this day.

DABRONA, Ire. a country or hilly parts on the division spring water part. Hence Dab and Dabble, the spring water part or place.

DACIA, on the Danube, the division spring water part confines. Dacre or Dacore, Cumb the division spring water border. Dactyli, Curetes or Corybantes, the division spring water side place, family or guardians. Hence Dace, the division spring water; Dactyle, of one long and two short feet or measure in musick or dancing; Dad, the side spring division; Daggle, the division spring water place; Dainy, the inclosing spring water part side; Dairy, on the spring water part; Daisy, the spring side.

DALE or Del, Scot. Wales, and North of England, the division spring water place, vale, or holding. Del-aley-boge-by-main-ington-keith-matia-rea or reudin, Salop, Scot. Leic. Cumb. Suff. the spring water place divisions, holdings, or vale, alley, border, dwelling part, inclosed or narrow part, confines town, water side, surrounding water side, or river side or Gaulish possessions or possessors. Bede considers the Dalreudin, the holders or possessors of the vales, on the river sides, as a third nation of Scots from Ireland. But the Welsh look upon all the Alcwyd Britons as far as Dunbarton, to be from the Cumberi, and Bede admits the lowland Scots to be of an English origin, and to me they seem to be a branch of the Cumber-

beri or Wenta Silurum and Belgarum, who went from Kent along Watling ftreet, or the Iter Cymbrorum firft into Wales, and from thence along the fame ftreet, which ran on the weftern fide of the ifland to Dunbarton or Britton. And no other ancient hiftorian mentioned any fuch colony, and the name Scot is a mere local appellative, fignifying the lower water confines or wood fide, and ancient writers meant no other than the Cumberi or Alewyd Britons, and Novantes or Wenta or vale dwellers; but was it true that thefe people came from Ireland, their anceftors muft have been before in Britain, and defcended from fome of our Britons. Hence Dale, a part or holding; Dalliance, on the fpring water place fide; Dally, on the fpring water place.

Damnii, Scot. on the furrounding fpring water part or vale. Dam and Damgarten, the furrounding or about the fpring water or confines water part fide. Hence Dam and Dame, the furrounding fpring water part without flowing; Damage, the furrounding fpring water part edge; Damalk, the furrounding fpring water part below its confines; Damn, on the furrounding fpring water part which was deemed the greateft flavery; Damp, the furrounding fpring water part; Damfel, the lower or leffer furrounding fpring water part.

Dan, Dania, Danum, Chefter, Denmark, and Doncafter, at the fide of or below the divifion fpring part or city. Danbury-by-denis-efend or Dacorum-monium-monith-tfey-trey or Wentrey-ube, Eff. Yorkf. Devon, Herts, Corn. Wales, Wilts, N-ham. on, below, or at the fide of the divifion fpring water burrow, dwelling part, lower vale, border, furrounding water or fea, mountain, lower fide, water part or Veneti water part town and divifion fpring water part. Hence Dance, on the fpring fide, or under the fpring of mufick; Dandle, on the fpring divifion water place fide; Danger, below the inclofing water place; Dangle, on the divifion fpring water place; Dank, on the divifion water confines.

Daphnis, the fon of Mercury, who plighted his faith to a nymph, the divifion fpring water part in the fide or lower part. Hence Dapper, the fpring water part; Dapple, the apple divifion.

Dardan or Troy, on the water part fide. Darent or Trent, Nott. the fide or divifion water part. Dar-ioritum-king-laftonley, Kent, Surry, Staff Der. the divifion fpring water part ford, confines, place fide town or dwelling place. Darnii or Antrim, Ire. on the divifion water part fide, rim or border. Hence Daring, on the inclofing water part or border; Dark, the divifion water part confines, which were covered with
wood

woods and thickets; Darling, the inclosing spring water place: Dart, the side spring water part; Dash, the spring part low; Date, the spring part or side.

DAVENPORT, Chef. the gate or port on the division spring water part. Daubeny, Surry, the division spring water part end. Dauphin, or Dolfin, France, on the division spring water part or vale side, edge or head, or a chief holder, like the English Adeling. Daurona or Cork, Ire. on the division spring water part border or confines. Dawtrey, Surry, the division spring water part or town. Hence Daub, the division spring water part; Daughter, the division spring inclosing water part; Dawn, on the spring; Day, the division or spring division.

DEA or Dee, N-Wales, Chef. Scot. the side or division water. Deal, Kent, the side or division water place. Dean forest, Gloc. on the side or division water forest. Hence Deacon, on the side or division water confines; Dead, Death, and Deaf, without spring; Deal, the side water place, the usual place of exchanging goods: Dear, the division or side water part; Dearth, the division water part side or aside.

DEBEN, Suff. on the division or side spring part. Debenham, a vale on the side spring part. Hence Debar, the side spring water part or bar; Debark, the side spring water part confines; Debase, below the side spring part; Debate, at the side spring part; Debauch, the side spring water part confines; Debt, the division water part, or a payment for the defence of the borders or country.

DECEBALUS, an appellative of office amongst the Daci, the spring water place covering border or keeper, as does Cassibellan signify, the spring water place or bank keeper; Bellinus, on the spring water place or bank; Cattimarus, the surrounding water part or meres keeper; Teutomarus and Teutobochus, the side spring meres or border keeper; and Cass or Gessi, the lower or lesser confines keepers or servants. Decombes or Decumanus, a supposed saint of Someriet, the side spring water confines port, haven or vale. He is said to have landed with one Carentocus a Briton, at Carenton in this county, from South Wales, probably alluding to some colonies of the Cumberi or Venta Silurum settling here. Hence Decade, dividing the inclosing water part; Decadency, the fall of its water; Decalogue, the ten laws; Decamp, without the inclosing water part or camp; Decant, dividing the inclosing water part on the side; Decay, without inclosing water or covering; Deceit, without sight; Deck, the side covering; Decree, dividing the inclosing water parts or borders.

DEMSTERS, the judges or magistrates of the Isle of Man,

the ifland divifion fpring water parts or diftricts, or the dividers or decreers thereof. Deepdale-den-ham-ing, Surry, Norf. the divifion fpring water vale, or a deep or dark vale or den in a fecondary fenfe. Hence Deed, the divifion part; Deep, the fide water part; Deer, on the fide fpring; Deface, dividing the fide part; Defect, without effect; Defence, the fide or divifion fence; Define, dividing the border; Degree, the confines or furrounding water divifion.

DEIRA or Deirland, that part of the kingdom of Northumland as lay fouth of the Tyne, comprehending Lanc. W-mor. Cumb. fouth of the fide water country; Bernicia, the other part of that kingdom, taking in all the other fide of that river to the Frith of Edenburrow, fignifying on the inclofing fpring water upper part. Hence Deject, and Deign, as from an high part low, like the funs rays on the earth, or from north to fouth.

DELGOVITIA or Wighton Yorkf. the inclofing fpring fide vale or town. Delta, a city of Syria, and an ifland of the Nile, the divifion fpring water place fides or poffeffions. Delvin, Ire. the divifion water or vale edge. Hence, Delay, the divifion fpring water place which occafioned delay; Delectable, at the divifion fpring water place able; Deliberate, at the divifion fpring water place; Delight, the divifion of light; Delineate, the divifion of the fide lines; Deliver, the divifion fpring water place; Delve, dividing the fpring water place; Deluge, the divifion fpring water place or rain covering.

DEMETIA, Dimeta, or South Wales, the furrounding water fides or vales, or the fouthern fides: thofe of Northumberland, or rather Northehumberland, as named in ancient writings, being the northern Meate, or dwellers in vales or meadows. Hence Demefne and Demean, on the furrounding water part fide or divifion; Demand, the fide part divifion; Demi, the fide or border divifion; Demife, below the border; Demon, on the furrounding water part fide; Demur, the furrounding fpring fide.

DENBIGH or Denby, Wales, the divifion water or vale part or habitation. Denchworth, Berks, the fide fpring water confines vale, or by the inclofing fpring water fide. Dengeneffe, Kent, the inclofing water vale promontory. Dengy, Ifl. the inclofing water fide or vale. Denmark, Dania, Jutland, or Gutland, Wetailaheed, or Cimbrica Cherfonefes, on, at the fide, or below the furrounding water or fea confines, or the fame as Dan in Brydan; the Danes having made ufe of the fame names as the Britains and Cimbri, and according to the fongs of the Britifh poets, the Druids with many followers

went

went to Denmark on their expulsion by the Romans ; or Dan as well as Tane or Thane, may signify on the inclosing water part side warden or lord. Hence Den, on the division water or vale; Denominate and Denote, the name or mark on the surrounding water part side; Dense, the inclosing water part side; Deny, no inclosing spring water part or entrance therein.

DEPTFORD, Kent, the dwelling water part side, way or ford. Hence Depart, dividing a part; Depend, deep end; Deprive, without the spring water part; Depth, the division water part side; Depute, at the side or division water part.

DERBY, the division water or dwelling part. Dereham, Norf. the dwelling water part vale. Derhurst, Gloc. the dwelling water part forest. Derlington, Dur a town on the inclosing water part place. Derry, Ire. on the division water part. Dert, the side water part. Derventio, or Derwent, Kent, Der. Yorkf. Dur. the inclosing spring sides, or Veneti water parts. Hence Derive, the flowing spring water part ; Derogate, the spring water part at the sea, which lessens or decreases.

DESMOND, Ire. the lower or farther side of or below Moun-ster, or lower surrounding water or sea side part. Hence Descent, on the lower inclosing water side; Desert, below the division water side; Design, on the inclosing water side or marks; Desire, the side spring part; Desist, the side spring lower part; Desperate, at the side spring water part; Despite, the side spring lower part ; Destroy, lessening the border wa-ter.

DETBURY, or Tetbury, Gloc. a spring or dwelling part or burrow at the side or division spring. Hence Detach, at the side spring water confines ; Detail, the side spring part flow-ing; Detain, on the side spring part; Deter, the division water part side; Detest, the division water part lower side or part ; Detort, dividing or destroying the border water side.

DEVA, Dewa, Deunana, Doncaster and Chester, the di-vision or inclosing spring water part or side. Deucaledonia, Scot. the southern or spring side inclosing water place banks or hills. Deveril, Wilts, the unseen spring or diving water place. Devi, S-Wales, the side spring; Devises or Devices, Wilts, the division spring sides or streets, or Wiccian confines, or Danith, as some imagined. Devon or Demon, Mon or Corn-wal side, deep part or vale. Dentburrow, Yorkf. the side or division spring side burrow. Hence Deuce, the division spring sides; Devest, the division spring lower side or uncovering ; Deviate, from the spring side; Device, the division spring side;

fide; Devil, the divifion fpring place; Devife, the dividing the fides or poffeffions; Devote or Demote, at the furrounding water part fide, places of devotion being furrounded with water or moted; Devolve, the rolling of poffeffions; Dew, the fide, furface covering, or dilated or divided fpring water; Dexterous, the inclofing water part divifion.

DIANA, faid to be the chafte goddefs of hunting, whofe temple was in Bithynia, the inclofing fpring water divifion or fide, or the fpring water inclofures or borders. Dictum, Diganwy, or Ganoc, N-Wales, the divifion or inclofing water fide ; Dicul or Culdee, the inclofing fpring water fides, divifions or dividers. Digby, Rut. Linc. the fide or divifion water part or habitation. Dillon, Ire. the divifion or fide water place bank or lane. Dilfton, N-hum. the fide or divifion water place fide town. Din, Dinan, or Dinas, Salop, Wales, Scot. the inclofing or inclofed fide or city. Dinder, Heref. the inclofing water part fide or hill. Dingle, or Smerwic, Ire. the inclofing or leffer furrounding water place fide or ftreet. Dingwal, the inclofing fpring water place fide. Dinnet head, at the inclofing fide head. Dis or Difs, Norf the lower fide. Difert, N-Wales, the lower water part fide. Ditch or Diche Marfh, Yorkf. the mere ditch or inclofing fide or part. Ditton, or Dichton, Surry, Camb. below on the divifion water fide. Divitiacus, chieftain of the Sueffiones of Belgic Gaul, the inclofing fpring water fide or Seine chieftain. Divodurum, the fide fpring water part gate. Divona, the fpring divifion or fource. Hence Dial, dividing place; Dialogue, dividing difcourfe; Diaper, dividing the parts; Dice, dividing the fides; Dictator, at the inclofing water fide man; Die, from or out of fpring or without; Dig and Dike, the dividing fide water; Dilacerate, dividing the water place at the inclofing water part or border; Dilate, dividing the fide place or parts; Dim, without the furrounding fpring; Din, or the fide fpring or ear; Dip, the fide fpring part; Dire, the divifion or dark water; Direct, the divifion water confines or border fide or poffeffions; Dirt, at the water part fide; Diforde, without, below or at the lower fide of the fpring, or water fpring place, or under the fide or ground; Difchain, without or below the inclofing water place fide; Difarm, without the arms; Dis or dith, the lower or under fide, or a part privative or without fpring; Difk, a fide or part that, that enlightened; Diffect, dividing the inclofing or furrounding fide or parts; Diftant, the fide below the fide or borders; Diftill the fide flowing part; Diftrict, the fides or leffer inclofing water part fides or poffeffions; Difunion, dividing or feparate inclofing union; Ditue, ufe, without ufe; Ditch, the fide inclofing water; Due, the fide fpring or unfeen; Divers and Divert, the fide fpring, or fpring water divifions; Diveft, the fide fpring lower fide; Divide, the fide fpring divifion; Divine, on the lower or future fpring divifion; Divorce, the border fpring fide divifion; Diurnal, on the furrounding fpring divifion; Dizen, inclofing the lower fide.

DOBHAM, Yorkf. the furrounding fpring water part or border, vale or home. Dobuni or Boduni, Gloc. Oxon. the inclofing fpring water part borders. Doctan, Scot. below or on the boundary water fide. Dodbrooke, Devon, the border fide brook. Dodington, Chef. a town on the boundary fpring water fide. Dol-gelley-wyddelen-zel, N-Wales, Scot. the cell, fpring fide bank or low vale. Donegal or Tir Connel, Ire. on the vale or Gaulifh borders, or the corner or inclofed water place. Done-main-maws, Ire. on the fea or fpring fide boundary or border. Dore, Dor-chefter-dan-manchefter-mernford, Heref. Dorfet, Oxon. Hunt. Bucks, the furrounding water part or border fide, gate, port, lower part, city, city part, and ford or way. Dove, Staff. Der. the boundary fpring. Dover, Dorober, or Dubris, Kent, the fpring water part on the boundary water, a harbour, haven, port or water gate. Dounamvenny, Gloc. the parts about the furrounding fpring water. Dow-ard-bridge-nham or Dunum, Yorkf. Heref. N-ham. Suff. Norf. Ire. the furrounding fpring or river part, fide, bridge, or vale. Hence Do and Done, inclofing or dividing the fide or boundary, and the boundary inclofed; Dock, a fhut or inclofed water part; Dog and Doge, the boundary or country chief; Dogma, over the country chief doctrine; Dolphin, the chief holder; Domain, the houfe or lands inclofure, or the chief furrounding border, or what is included within a man's own borders, as of his own original right, without the gift of another; Domefday, the divifion of the kingdom; Don, on the boundary; Door, from the border; Dorn's Pence, on the furrounding water part gate pence, or a coinage of brafs and filver, for the difcharge of portage at Dorchefter, which was commonly called king Dron's pence; Dot, the furrounding part; Double, the fpring divifion water place; Doubt, the fpring divifion part; Dove, a tame or under covering animal; Dower, the border divifion; Down, on the inclofing fpring water, which is low or down; Doxy, the boundary water confines or companion.

DRAICOT, Staff. the inclofing water part fide or wood, as being on the border fide. Draitton, Salop, the water part fide town. Dramore, Ire. the fea or furrounding water part
or

or moor, if it happened to be a moor as was commonly the cafe on the borders. Hence Drab or Draff, a divided fpring water part; Draft, a divided water part fide; Drag, Draw, Dray, and Drake, the fide water part confines; Dram, the furrounding water part; Dreadful, the fide water part full; Dream, the fide furrounding parts; Drear, on the fide parts; Drain, the inclofing water part; Drefs, inclofing the lower part or covering.

DRIBY, Linc. the fide water part or habitation. Driffield, Yorkf. the fide water part dwelling place. Drighlington, Yorkf. a town on the inclofing water part fide. Drimen, Scot. on the fide water part border. Drimlar, Scot. a place on border fide. Dripoole, Yorkf. the fide water part poole, lake, or furrounded water place. Hence Dribble, the fide water part flowing place; Drink, inclofing the water; Drip, the high water part; Drive, the fpring water part; Drivel, the fide fpring flowing part; Drizzle, the fide water part place, which are the properties and accidents of water places.

DROGHEDA, Ire. along the furrounding water part. Droitwich, Worc. the furrounding water part fide confines or ftreet. Hence Droll, on the furrounding water place; Drone, on the furrounding water part; Drop, a furrounding water part; Drofs, the furrounding water part fide; Drought, the furrounding water part fhut; Drown, in the furrounding water part; Drowfy, at the fide of the furrounding fpring water part.

DRUIDES and Dryades their nymphs, the divifion fpring water part fides or dividers, or the chieftains or heads of clans; As the ftory told by Parthenius, of a chieftain of the Cavares on the bank of the Rhofnes facrificing the wife of Xanthus, for her criminality towards her hufband, which no body but a druid could do, feems to confirm. Drum-albin-bough-laurig-ond, Scot. Cumb. the furrounding fpring water on the hill parts, confines part, water bank, and on the fide or mountain as it happened. Drury, Norf. on the fpring water part. Drufomagus, Gaul, the furrounding divifion, or middle fpring water confines gate. Drylton, Scot. a town on the fpring water part. Dryftoke, Leic. the without water fide ftation. Hence Drub, the fpring water part, which beats againft the fea; Drudge, the fide fpring water part edge; Druor, the furrounding water, fea or founding part; Dry, wet part without water.

DUBLIN, Dublin, Eblana, or Bala Cleigh, Ire. the divifion fpring water place, poole or vale, which the Irifh render a town of hurdles, as having its foundation of hurdles, but moft likely from its being o iginally built of hurdles, watles, or wickers, like Hen. II. palace there built of Watles, after the manner of the country, Wicklow and Tu Gwin, the princes palace in Wales. Duck, the spring water part confines. Dudden, Lanc. the division spring side den or vale. Dudley, Staff. the division spring side place or dwelling place Dugdale, the division spring water confines vale. Duglas, Lanc. Scot. the division spring water confines place, or green place. Dui, Wales, the division spring or its sides. Dulwich, Surry, the division spring water place confines, street or fortified part. Dum-barton or britton-blane-norix-frise-na, Scot. Gaul, the furrounding part, bar or British town, bank, chieftain, side or lower part, and a part. Dun or Dunun, on the division spring bank, hill or town. Dun-bar-barton-canon-dalk-dafs-dee-eaden-efty-evet-fermling-garvan glas hill-keld-keron-keten-kirk-loch-macus or magu-mow-ing-etyr-robin-foot-fany-flaburgh-ftafag-ftable-ftaville-ter-winden, Scot. Ire. Yorkf. Corn. Wilts, Flanders, En' Berks, Suff. Leic. N-hum. Bedf. Devon, on the division spring part or bar, bar town, border, water place confines, lower part or side, side, water den. vale or town, lower side, whole, furrounding water part bank, water part confines. confines place, hill, including water place or ell side, border or water border, confines, town, lake, chieftain, fortified hill, inclofure, water side, border part, the lower inclofing water or for den, on the fea or lower fide flation burrow, water part inflow, bank or hill flation place, inns or ftables, flation dwelling place or village, ell water part, ftreet inclofed or fortified, and inclofing water fide vale. Duplin, Scot. the division spring part bank. Dor-den cee, Ocobrive, Wenmer or Reburn ham-inclofe-Severin or Dorchefter, Obernium or Canterbury-obrevis, Dernford or Rochefter, ocafs-ocoterum-olenum or Lenham-Spontes or Germanchefter-Otrige, Dorotiges or Derks, Heref. Wales, Herts, Dur. Scot. Dorfet, Kent, Hunt. the division spring water vale or fide, mere, borders or Veneti mere, vale, in the lower fide, on the fea or haven, dwelling part or Kentish dwelling part, way, ford, city or inclofed place, lower fide, including border, bank, furrounding fides, parts or of the Wapontache, and on the water part or fo fide or gut part. Dutch, Teutch, or Tutico, the fide or division lower confines or country, its prefent Belgic or Batavian inhabitants being always fated on the Rhine. Dutton, Chef. the division spring fide town. Dyvi, Wales, the division spring water. Dynevor, a caftle on the top of a hill call from Carmarthen, reputed to have been the chief refidence of the princes of

South Wales, a hill or fortified place, or city on the furrounding water part. Hence Dual, the divifion fpring water place;
Dub, the divifion fpring part; Dubious, dividing the fpring
part; Ducal, on the divifion fpring confines; Due, from the
divifion fpring; Dug, the fide fpring water fhut; Dull, on
the divifion fpring water place; Dumb, furrounding the divifion fpring part; Dun, the divifion fpring one or man;
Dupe, dividing the f ring part; Dur, the divifion fpring, or
the fpring water; Duft, the lower fide divifions; Duty, the
divifion fpring fide; Dwell, on the divifion fpring water
place; Dye, without fpring; Dynafty, the government divifion.

EADBURG, Kent, the water fide fpring or dwelling part or
burrow. Ealandon, War. on the water bank or lane fide or
land. Earles-dike-crofs-town, Yorkf. Scot. the inclofing water place, or earles dike, crofs and town. Thirlftan, the inclofing water fide or earles town. Earles Coln, on the earles
inclofing fpring water place. Earles dike, near Bridlington,
was made by the earles of Holdernefs, as a boundary betwixt
t e Gabrantovici or Brigantes and the Pontes, Wapontaches,
or Coritani, where the earl prefided, as did the lord over the
Lair Merney in Effex, till they were made merely official, by
Alfred's vefting their power in a vifcount or fheriff of his own
appointment, and the lords power of courfe in himfelf. Earle
feems to be a tranfpofition of Lares, Lars, Lards or Lords, the
fons of Lara by Mercury, fignifying over the fpring water
places, or the river boundaries, or the heads of the Tufcan
clans or tribes, like the bards, barons, or knights amongft the
Britons, and alderman over the fea-faring and mercantile parts.
Eaft Anglia, the fea or water fide corner or angle, or the
eaftern corner from its fituation in a fecondary fenfe. Eaft-
bourn, Suff. the fea fide neighbourhood. Eafter, at the fea
fide feafon. Eafterford, or Eaft Sturford, Eff. the water fide
ford or way. Eafterling, or Sterling, Scot. the lowland inclofing water part or confines. Eaftonefs, Suff. the lower
water fide border promontory. Eaft Mean Vari, Hants, on
the fea fide middle or divifion fpring water place, as Weft
Mean is the inclofing fpring lower fide part. Eaton, Berks,
on the border water. Eaugh, Ire. the confined fpring water.
Eay, Suff. the water. Hence Each and Eager, the water or
river fides or confines; Ear, on the fpring; Earldom, the
earles limits; Early, the fpring water place; Earn, on the
fpring; Earnett, on the fpring lower fide; Earth, the water
fides, divifions, parts or poffeffions; Eafe, the water fide,
lowest

lowest place, or on the fide; East, the lower water or fea fide; Eat, at the fpring water; Eaves, the fide water part.

EBBA, or Ebbes fleet, Kent, the flowing fpring water part or flood. Ebburton, Gloc. the flowing fpring water part town. Ebchefter, Dur. the fpring water part city. Ebercurney, a monaftery on the Tyne, whereof Trumwyn was abbot according to Bede, on the border fpring water part. Eblana, Ire. the fpring water part bank or vale. Eboracum, or York, the border fpring water part confines or city. Eburovices, on the river Ure, at Eureux in Normandy, the fpring water part confines or ftreets. Eburones, near the Ourt in the Netherlands, on the fpring water parts or the burrows. Hence Ebb, the flowing water part; Ebriety, the flowing water part fide; Ebulition, on the flowing fpring water part fide.

ECCLESHALL, Staff. on the inclofing water place, or a hall thereon. Hence Eccentric, the inclofing water part confines, which is from the center; Ecclefiaftic, the inclofing water fide confines or covering; Eclat, at the inclofing water place; Eclipfe, the feeing part inclofed or fhut; Eclogue, the country difcourfe, Economy, the fhutting the furrounding parts; Extacy, the lower inclofing water fide confines.

EDELINGHAM, N-hum. the inclofing water place or holding vale or home; and Adeling, or Edeling, the heir apparent, he was fo called from his being the holder or defender of thofe boundaries, or of England, or Ing-land, the kingdom land or boundaries, or Ci-ing or Kyngdom, the limits of the Ing, Kingdom or England chief.

EDEN or Eaden, and Eden-burrow, W-mor. Scot. Syria, Cilicia, and other parts on the Mediterranean coafts, and the garden of Adam, the inclofing divifion of earth and water, the divifion fpring water parts, or the vale and the vale burrow. Edern, N-Wales, on the water part. Ederington, Suff. a town on the inclofing water part. Edgcomb, Corn. the water edge vale. Edge hill, War. the border water or edge hill. Edgeware, or Edgworth, Midd. the fide fpring water or border edge. Edington, or Eathandun, Wilts, the inclofing water fide or edge town. Eden, in Phrygia and Thrace, on the divifion fpring water part or vale. Hence Eddy, the double or divided water fpring; Edge, the water fide or confines; Edify, the water fide part; Edition, on the water fide; Educate, at the divifion fpring confines, the feat of learning. Moft of the many attempts made towards a difcovery of the fituation of the paradifal garden, fix it in the neighbourhood of the Mediterranean. To me it feems probable that it is al oppofite the Nile, in the Mediterranen vale, which has been fince covered

covered by that fea, as have been many others by the feas and
gulphs that were all formed by deluges, except the ocean which
furrounded the continents of Afia, Europe, and Africa. For the
prophet Ezekiel, where he fpeaks of the merchants of Eden
trading to Tyre, probably meant thofe of the Mediterranean
fea ; and Mofes, in faying that the garden of Eden was planted
eaftward of Eden, and ftocked with all forts and kinds of trees,
vegetables, and animals, muft mean a more extenfive fpot
than has hitherto been affigned, for a garden that was to produce, fupport, and ferve for the increafe of fo many things,
and had the Euphrates, the elevated and divided Nile, and
two more large rivers for the watering or overflowing of it.
But it feems moft probable that Eden or the vale was the ground
now covered by the Mediterranean fea, and that the river
which went out of Eden to water or overflow the garden, that
was parted and had four heads or fources, were the Tanais, the
Nile, and the Euphrates, above and below the garden, which
eaftward of the garden was fo called, as being the fpring part
fide, and after it paft the garden weftward towards the fea
Hiddekel, the vale length water to the fea ; and the Pifon encompaffing the land of Havilah, fignifies on the lower or fun
fetting part or Europe, the border of the upper or eaftern
dwelling place, as does Gihon which encompaffed the land of
Æthiopia, or the fun fide part fignify, inclofing the fun fide,
or dividing Africa from Afia ; and amber, gum, aga, and
gold have been found in the northern parts, not far from the
Tanais. And if the deluge did not form thofe feas, and divert
the courfe of the Tanais and Euphrates, though it is faid that
the Euphrates once went a different courfe, the Tanais might
mean the Euxine and other feas to the Mediterranean, where
might be the borders of the garden, but it feems more probable
that the Nile joining the Tanais, was its lower or weftern border, that the Euphrates running from the eaft divided, fo as
to form a Δ or delta, the figure 4, a model or compound-

ed of four Y's thus put together (Δ / Y) and which courfe has

been obferved in the future divifions of countries by rivers,
motes, or commotes, and which divifion ftill appears on the
divifion of the Nile below Cairo, and its elevated fituation
above the plains, different from all others to overflow the
country is remarkable.

EFFINGHAM, Surry, the ebbing and flowing fpring water
part vale or home. Hence Effable, howards Effacc, leffining
the flowing fpring part; Effect, the flowing fpring conti-
nues ftill ;

division; Effeminate, the flowing spring part at the surrounding water part or sea side; Effervescence, the flowing spring water part springing in the lower confines; Efficacy, the flowing spring part inclosing side; Efflux, the flowing spring water place; Effort, the flowing spring part for the side or border; Effusion, the flowing spring part on the side.

EGEON, Greece, on the water edge. Egleston or Eglinton, Dur. Scot. a town on the water place side or bank. Egremont, Cumb. the surrounding water or sea side border. Hence Egg, the inclosing spring source, seed or part; Egistment, the inclosing spring lower side part, or commons let out for the egistment of cattle, Egregious, the inclosing water part confines; Egress, below or at the side of the border or inclosing water part.

EIGHT, Alney, or Oleanag, an island in the Severn, the inclosing water place part or bank, or an island, or the side land, as being twice four, the surrounding sides or parts. Eimot or Emot, W-mor. Cumb. the side surrounding or rim spring part, or the mote. Eira, Scot. the water part. Hence Ejaculate and Eject, at the inclosing spring water place edge; Eigh, the inclosing water edge; Eighth, the eight; Eighteen, eight on the side of ten, tyes in or inclosures; Eighty, eight ties, sides, divisions or districts of eight possession, which, with the two possessions or sides of the island in the boundary water, made up ten or the tything, or the inclosing water tye; Either, the side spring water part, which may belong to either side.

ELAN, Alan, Eland, or Elandun, Wilts, Suff. Norf. Heref. Cumb. N-Wales, on the water or spring water place, side, bank, land or town. Eleutheros, Eleutherius, or Eleutheria, a river near mount Libanus, and the names of Jupiter and his feasts, the flowing spring water place. Eleusis, near Athens, the spring water place side, where the Eleusinian mysteries, instituted by Ceres, or the hieroglyphicks were exhibited, and to which, as being a stranger, Hercules was refused admittance, as the Greeks pretend, but it seems more probable that the Greeks were therein instructed by Hercules. Elford, Staff. the water place way or ford. Elgin, Scot. on the inclosing water place. Elham, Kent, the water place vale. Ellsmer, Salop, the mere side water place, or the border palace. Ellestre, Herts, the high or hill place town. Ellingham, Hants, the inclosing water place vale. Emesley, Elmley, Elmet, or Ulmetum, Worc. Yorkf. the middle, division or surrounding water place side spring water or dwelling place. Elham, Norf. the surrounding or sea water place vale or marsh. Elmore, Gloc. the sea surrounding water place. Elfing, Norf.
the

the inclofing water place. Elftow, Bedf. the water place fide border or ftation. Eltham, and Elton, Kent, Hunt. the water place fide vale or town. Elwy, N-Wales, the fpring falt water place. Ely, Camb. the water place, falt water place, or the ifland. Hence Elaborate, at the water place border; Elepfe, the water place fide; Elaftic, the water place lower fide confines; Elate, at the fpring water place; Elbow, the water furrounding place or bow; Elder, the water place fide; Elder, or Alderman, the water place fide man; Eldeft, the water place loweft fide; Elect, the inclofing water place fide; Elegy, on the water place inclofing, the ufual burying places of our maritime anceftors; Element, the furrounding place or water in the fides, or on all fides; Eleven, above the inclofing fpring water place, or ten; Elf, the fpring water place being; Eligible, the water place confines able; Elifion, on the water place fide; Ell, the water place, or the diftance betwixt a perfon's legs when both are extended, which is the water place of man; Elope, from the inclofing water place or border; Elfe, the water place fides; Elufion, and Elyfian, on the fpring water place lower fide.

EMBRUX, in Narbon Gaul, on the furrounding fpring water part. Ember days, the furrounding fpring part divifion. Emely, Ire. the furrounding water or its margin. Emildon, N-hum the water place margin town. Emlin, S-Wales, the furrounding water place bank. Emporium, a town of Celtica, or open ports where public fairs were held by confent of all nations, fubject only to the controll of their own magiftrates, as Genoa and London, and its precincts, including the river Thames. Hence Emaciate, the furrounding fides together; Emanation, the furrounding fpring on the fides; Emancipate, together at the furrounding confines part; Embalm, covering in the furrounding water place; Embark, the furrounding water part fhut; Embay, the furrounding water part; Embrace, furrounding the fpring water lower part, as the fea does the river; Emerge, from the fea edge or fide; Eminent, on the border fide; Emit, the furrounding water fide, like a river into the fea; Empire, the furrounding water part; Employ, the furrounding water place; Empty, the furrounding parts fides; Emulate, at the furrounding fpring water place fides.

ENDERBY, Bedf. the inclofing fpring water part, or habitation. Enfield, Midd. the inclofing or border dwelling place. Engain, N-ham on the inclofing water place. Engelbert or Engelbard, the inclofing water place or fide part or bar. Engerfton, Eff. below or at the fide of the inclofing water fide.

Engelham,

Engesham or Ensham, Oxon, the incloſing water ſide vale. Englefield or Inglefield, Berkf. the water incloſing iſland or Engliſh dwelling place ; it being a part incloſed by the Thames and Kennet and a ditch cut from one river to the other. Engleſeria, the water incloſing or Engliſh borders. England, Ingland, Anglia, Angliterra, Engla, Engle-lond, Angleynne, Engleynne, and Ing-le-iſh, even in the time of the Heptarchy, the water incloſing, Angles or Iſle land, Kingdom or Ki-ing or Kyngdome ; Angles being a tranſlation of Iſceni, the corner Iſland ; as is Angleſea, the ſea Iſland and Ing in the Ancient Britiſh dialect of the Celtic an appellative expreſſive for a part wholly incloſed by water or otherwiſe, as are Ang and Eng of a corner ſo incloſed, and Ong of a place wholly covered by water or otherwiſe, as appears under Ang, Eng, Ing, and Ong. Hence Enable, incloſing the ſpring water place ; Enchant, on the incloſing water ſide ; Encloſe, a locking or ſhutting in the ſide water place ; Encounter and Encourage, on the border edge or ſide ; Encroach, on the border confines ; Encumber, on the ſurrounding ſpring water part ; End, the incloſing ſide ; Endleſs, the place below the incloſing ſide ; Endorſe, below the incloſing ſide border ; Endow, the incloſing ſide ſpring circle or border ; Endue, the inſide ſpring ; Endure, the incloſing ſide ſpring ; Endwiſe, the incloſing ſide ſpring from below ; Enemy, the ſurrounding one ; Enfeoff, incloſing a part off ; Enfilade, an incloſed ſide place part ; Enforce, in or on the ſurrounding water part ſide ; Enfranchiſe, within the incloſing water part confines ſide ; Engage, the incloſing water edge ; Engliſh, the water incloſing place or Iſland iſſue ; Engrave, the water incloſing border parts, the prototype of engraving, writing, and engroſſing ; Enjoy, in the circle ; Enlarge, incloſing the place on the confines ; Enough, the ſpring water circle incloſed ; Enrol, on the border water place ; Enſign, the ſign one ; Enſue, on the ſpring ; Entail, on the ſide place or race ; Entangle, on the ſide Angle ; Enter, in the ſide water part ; Entice, the lower or female ſide ; Entire, the incloſing water ſide ; Entity, a part or ſpring within the ſides or covering ; Entry, in the ſide water part ; Envy, in the way ; Envoy, on the way.

EPIDAMNUM, Dyrrhacium or Durazzo in Macedonia, on the ſurrounding water dam or ſtopped up part. Epidium or Epidion, one of the Hebrides, or weſtern Iſles of Scotland, the weſtern water ſide part. Epping Foreſt, Eſſ. the incloſing part foreſt. Epſom or Ebſham, Surry, the ſpring water part ſide vale or dwelling part. Hence Epact, beſides the incloſing ſide ; Epic, on the confines or country part ; Epilogue, an after diſcourſe ;

courfe; Epifode, the part below the ode; Epiftle, the part below the ftile; Epitaph, befide the covering part; Epithet, the part at the fide; Epitome, the furrounding borders part; Epocha, the furrounding part fhut.

ERAUGH, Ire. the water inclofed country or promontory. Erdburrow, Leic. the fide water burrow. Erdefwick, Staff. the fide water, ftreet, fortification, or city. Erdini or Fermanagh, Ire. on the fide or furrounding water parts confines. Erdeley, Heref. the lower fide water place. Erebus, a river of hell, the fpring water part. Erefby, Linc. the fide water part or habitation. Eridge, Suff. the water edge or ford. Erlfton, Scot. the water place fide or carles town. Erming or Ermin Street, through Camb. Hunt. Norf. Suff. on the furrounding water fide or edge ftreet; or as it went from Southampton to Monmouth or St. David's, or from the fouth to Carlifle; it may, in a remoter fenfe, fignify the mine ftreet, as mines and ores were originally found along the fea coafts, from whence their names are derived. Ern, Scot. the inclofing high or difmal water. Ero and Leander, the fea and river. Eryth or Erytheia, Kent, Camb. Ionia, the water fide way, paffage or ford. Erwafh, Nott. the water wafh or low fpring waters. Erwy Porth or Erw'r Porth, N-Wales, the fpring water gate paffage part or port. Hence Era and Ere, the fpring or fpring water courfe; Erect, fpringing up the fides or parts; Eremite or Hermit, at the water border fide; Err, on the water; Errand and Errant, on the inclofing water fide; Erft, the fpring lower fide; Erudition, on the fpring water fide part; Eruption, the water up the fide or furface.

ESCOURT, Wilts, the lower inclofing water or border fide or court. Efcricke, Yorkf. the lower inclofing water or border fide; or creek. Efcuage, the lower inclofing fpring edge; Efquire or lower horfeman or knight. Efk and Effex, Scot. Cumb. Eff. the inclofing water fide. Efchequer or Scaccarium, over the tax or efcuage. Efcheate, below the inclofing water fide. Eflington, N-hum. the inclofing water place fide town. Effenden or Efton, Rut. vale, home or town on the fide. Eftrey, below the fide water. Hence Efcape, below the fhut part; Efchar, on the lower covering or border; Efcort, the lower border fide; Efpecial, on the lower confines part; Efpy, the lower or fide part; Effay, the leffer or lower attempt; Eifence, below the feen; Eftrange, below the inclofing water part fide; Efterling, the lowland inclofing water place, or a coin thereof called in contradiftinction of the highlands; Eftoppels, below the border water parts or out of poffeffion;

Eftovers,

Eftovers, fprings of wood below the borders, which belonged to the prince or lord ; Eftuary, water lower fide.

EiOCETUM or Wall, Staff. the border water fide or the wall. Etruria or Tufcia, Italy, on the fide or lower water. Hence Eternal, on the furrounding fpring fide ; Ether, the covering or fky water ; Ethic, on the poffeffion or country ; Etymology, a difcourfe on the furrounding fprings or words, which are the fprings of truth. Etymology no longer ftands on fo precarious a footing as analogy or mere conjectures, but is reduced to as much certainty as any other fcience, and may, perhaps, ferve as the criterion of truth.

EVANDER, an Arcadian, who drove the Aborigines out of Italy, and built a town where Rome now ftands, the inclofing fpring water part or the Tiber. Evenlade, Oxon, the inclofing fpring water place fide or country. Eula, and Euboea, Scot. Greece, the inclofing fpring water part. Evelthot or Eve-fhot, Dorfet, the fhut fpring water place. Evel and Evelmouth, Som. the fpring water place or mouth. Everley and Eburlegh, Wilts, the fpring water part place or dwelling place. Everfdon, Herts, the fpring water fide town. Evefham or Eufham, Worc. the fpring fide or fide fpring vale. Eumer, Yorkf. the furrounding water or mere fpring. Euphrates, a river, rifing in Armenia, the fpring water part fide. Eure, Euer or Wer, Bucks, the fpring water. Eufdale, Scot. the fpring fide vale. Eufdon, Suff. the fpring fide town. Euxine fea, the inclofing water or fpring water. Ewe and Ewel, Devon, Kent, Normandy, the fpring or fpring water place. Ewelme, Oxon. the furrounding fpring water place. Ewias, Heref the fpring or Wey fide or valley. Hence Evacuate, from the inclofing fpring water fide ; Evade, from the fpring fide ; Evaporate, a fpring from the border fide ; Eve or Vigil, the fpringing infide ; Even, the eve or fpring in or fhut ; Even. the fpring in the fides or poffeffions ; Ever and Every, the fpring ; Eugh, the high fpring or a tree ; Evil, the fpring water place ; Eulogy, the furrounding fpring water place confines or difcourfe ; Eunuch, one without the animal fpring ; Euphony, the furrounding fpring part ; Ewe, the fpring of animal or the animal ; Ewer, the fpring water ; Ewage, the fpring water edge or confines, or a toll paid for the liberty of a paffage or waterage.

Ex, Ecg, Ecs, Devon, the fide or lower water. Excefter, Exceter, or Ecgefter, the fide or lower water part confines, inclofed fortification or city. Exminfter, the fide water monaftery Exmore, the fide water more or furrounding part. Hence Exact, the inclofing water or border act ; Exalt, on the fide

fide water; Examine, the fide water edge or confines; Example, the furrounding fide water place, as a model or pattern; Exafperate, at the lower fide water part; Except, out of the fide water part; Exchange, the fide water confines or fhut; Excife, the fide water lower confines; Exile, out of or from the fide water place; Exift, out of the lower fide or furface; Exit, outfide; Exotic, out of the inclofing fide or country; Expatfe, out at the inclofing part fides or extended; Expel, out of the confines water place; Expert, at the fide water fpring part; Expofe, below the fide water part; Exprefs, the fide fpring water part below or at the fea fide, or the fpring out at the fea fide; Extend or Extent, out on the fides; External, on the inclofing water part outfide; Extreme, the furrounding water outfide.

Eya or Eye, Eff. Suff. the fpring or fpring water. Eymouth-ftoney-nfham-thorp, Scot. Suff. Linc. Bucks, the fpring water mouth, lower fide, inclofing fide vale or water gate. Hence Eye, the fpring of fight; Eyre, over the fpring either of wood or water.

FABERIA, an ifland in the German fea, the furrounding water part part. Fabaris, Italy, the fide or lower fpring water part. Fair-forland, Ire. the fpring water part foreland. Fair-ford, Gloc. the fpring water part ford or way. Fakenham, Norf. a vale on the inclofing fpring water part. Faleria, Tufcany, on the fpring water place or vale part. Falkirk, Scot. the fpring water place or vale border, fhut, inclofure, fortification, meeting place, church or city. Falkland, Scot. the inclofing fpring water place, bank fide or land Falkefley, Staff. the inclofing fpring water place, fide place, leet or palace. Falmouth, Corn. the fpring water place mouth. Fanelham, Germany, the inclofing fpring water place vale. Fanum, Germany, Italy, Greece, the inclofing fpring part or a temple. Fare, Farendon, and Farley, Staff. Berks, Som. the fpring water part, a town on the fpring water part, and the fpring water part place. Farnington, Gloc. a town on the furrounding fpring water part edge Farn, and Farnham, N-hum. Suff. Bucks, Surry, the inclofing fpring water part or vale. Faftcaftle, the lower or fpring water part fide caftle. Fauburn, Eff. a dwelling on the fpring part. Faunus the fon of Picus, and father of Latinus, the third king of the Aborigines of Italy, and Fauna, an Italian goddefs, the inclofing fpring part. Fawey, and Fawfeley, Corn. N-ham. the fpring water part, and the fpring water part fide place. Fax, Yorkf. the inclofing fpring water part. Hence Fable, on the fpring water place; Fabrick, an inclofing or brick part; Face, the fide part;

part; Factor, the inclofing fide or acting man; Fail, from flowing or fpringing; Faint, the inclofing fpring part fide, or off an infide; Fair, the fpring part; Falfe, below or at the fide of the fpring water place; Fame, the furrounding fpring part; Family, the furrounding fpring place or race; Fan, on the fpring part, Fane, Fanatick, and Fancy, the fan or fpring of wind part; Far, on the fpring water part; Farther, on the fpring water part fide; Fartheft, below or at the fide of the fpring water part fide; Fare, on the fpring water part; Farewel, dwell on the fpring or growth; Farm, the furrounding fpring water or inclofed part; Farthing, Feoling, Fardingdeal, Fardel, Farundel, Quadrantata, or a rood, the fpring water part parting inclofure, divifion, holding or rent; of which two made an oblota, halfpenny, or two reeds, and two of thefe made a denariata, an acre or a penny; 12 pence, or acres, made a fhilling, or folidata; 20 fhillings, or 240 pence or acres made a librata terra, or rent; 40 fhillings, or 480 pence made a military fee, maenols, or manors; 12 manors and 2 villes made a commot, and 2 commots made a canton, burrough, ftreet, or 100 poffeffions, military fees, or freeholds each, according to fome admenfurations, but others make it fomewhat more, befides waftes and commons, which were divided into townfhips, tythings, or tens, and thofe again fubdivided into villages, farms, and cottages. There were alfo farthings, halfpennies, and pennies of gold, anfwering the quarter, half, and a whole noble, or 6 fhillings and eight pence; Faft, a part below a covering; Fat, at the fide part; Fatal, on the fpring fide part; Farther, the fide or divifion fpring part; Faufet, the fide fpring water part; Favour, the border fpring part; Fawn, the inclofing fpring part.

FECKENHAM, Worc. the vale or foreft on the inclofing fpring water part. Felden, War. the fpring water place fide, fen or vale. Fells, Lanc. the low fpring water places or fens. Felding, N-ham. the inclofing fpring water place fide. Feld and Feltweil, Norf. the fpring water place fide or dwelling. Fern Ditton, Camb. a fen on the fide part. Fenten Gollan, Corn. a fen below the inclofing water bank. Fenwick, N-hum. the inclofing fpring part confines, ftreet, or fortified part. Ferbille, Ire. the fpring water part dwelling place or village. Ferden, Germany, on the fpring water part fide. Feringham, Suff. the inclofing fpring water part vale. Fermanagh, Ire. the inclofed water or lake parts. Fermoy, Ire. the fea or furrounding water part. Ferneby, Lanc. the inclofing fpring water part or habitation. Fertulogh, Ire. the lake fide water part. Feftiniog, N-Wales, the fpring part fide below the fea.

Fefcenium,

Fefcenium, Tufcany, the lower inclofing or confines water part. Fetherfton, and Fetherfton-haugh, Staff. Cumb. the high fpring water fide town. Fetiplace, Som. Berks, the fide fpring part place. Feverfham, Kent, the fpring water part fide vale. Fewes, Ire. the fpring water parts, fides, fens, or woods. Hence Fealty, the fpring water place fide, or the country defence; Fear, on the water part; Feaft, the water part lower fide; Feat, at the fpring water part; Feather, the fide fpring part; Feature, on the fpring water part fide, or the fhadow in the water; February, on the fpring part; Feculent, the fpring water confines bank fide; Fecundity, on the fpring water confines fide; Federal, on the fide fpring water part or border; Fee, off anothers fpring part or border, allodium being a member of the ancient poffeffons; Feed, of the fpring part divifion; Feel, on the fide fpring part; Fell, on the fpring water place; Felicity, the fpring water place confines fide; Feion, or Fe-cloin, the removing below the borders or ravifhing, contrary to Fealty and engagement of frank pledge; Female, the furrounding fpring part; Fen, an inclofed or fhut fpring water part; Fence, a fide inclofing part: Fend, a fide inclofing part; Feod, or Feoff, the furrounding fpring water part divifion; Ferine, and Ferocious, in the foreft; Ferment, the furrounding on the fides; Ferry, on the fpring water part; Fertile, the fpring water part fide place; Fervent, and Fervid, at the fide fpring water part.

Fife, or Five, Scot. the fpring water or dwelling parts, manors, or divifions. File, or Filey, Yorkf. Lanc. the fpring water part or dwelling place. Finborrow, Chef. the inclofing fpring, edge, or border burrow. Finchale, or Finkley, Dur. the inclofing fpring part confines place. Finchinfield, Eff. the inclofing fpring water part dwelling place. Fingale, Ire. the inclofing fpring water vale, Gaulifh place, Welfh, Englifh, and the ftrangers place; the Irifh, who were of the ancient Britifh or fea coafting race, confidered the whole Eng-lifh and Gauls, or dwellers in vales, as ftrangers, as did the old Britons of Scotland and England, the Novantes and Tri-nobantes, or Sulloniacæ, who were of the Gaulifh race, and dwellers on the fpring fides. Finerfin, Scot. the inclofing fpring water part rim or border. Firth of Forth, Scot. the fide fpring water part, ford, way, or paffage. Fifcard, S-Wales, the fpring or dwelling part below the inclofing water fide. Fifhpoole-ftreet, Herts, the ftreet or dwelling part below the pool, or furrounding water place. Hence Fib, from the fpring or truth; Fibre, the fpring of life part; Fickle, the upper fpring place; Fiddle, the fide fpring place; Fief, the fpring
dwelling

dwelling or living part; Field, the spring water place side or
dwelling part; Fiend, the inclosing spring side; Fierce, the
spring water part confines or forest; Fifteen, the dwelling part
and the inclosing sides and covering, or the sky or a side every
way; Fifth, the dwelling part divisions; Fifty, five inclosing
sides or tyes; Fig, the spring part shut or bag; Figure, the
inclosing spring water part, side or border, or the simple parts,
divisions or sides of forms, systems, or modes of things or names
in their combined and multiform appearances or existences, or
certain longitudinal sides, figures, or fibres, forming more
palpable bodies, as an assemblage of figures called a printer's
form, the forum, a formula, formedon, form of a writ, and
forma pauperis, not firum, firmula, firmedon, figure of a writ,
or of a poor suitor's proceeding in a law court, also the distinctions observed through all our definitions betwixt the O
and I, but their physical meaning must be left to the farther
discussion of physiologists, whose attempts have hitherto proved
fruitless; the names make no distinctions betwixt the qualities
of figures and forms, but as they merely affect the senses by the
different configuration of parts, as essences seem to be unde-
finable; File, the spring water part place; Fill and Filly, on
the spring water part place; Film, surrounding the spring water place; Filter, and Filth, the spring water place side; Fin,
the surrounding edge spring part; Finger, the border spring
part; Finish, the lower edge; Fir, the high spring; Fire,
the high spring; Firm, the surrounding spring part; Firma-
ment, the high surrounding spring part sides or covering part;
First, the high spring part to the lower sides or possessions;
Fish, the lower spring parts or heirs; Fit, the side spring part;
Five, the spring dwelling part or the manor; Fix, the spring
part inclosed.

FLADBURY, Worc. the side spring water place dwelling
part or burrow. Flamborough, Yorks. the surrounding water
place burrow or neighbourhood. Flamsted, Herts, Kent, the
surrounding spring water place side or station. Flanders, on
the inclosing spring water place side, bank, bank side or land.
Flatholm island, in the Severn, the side spring water place
island. Flavia Cæsariensis, the spring water place way or street
of the Isceni. Flavia Æduorum, the spring water place way,
or street of the migration of the Ædui, a clear admission
of the Romans themselves, that the principal streets were the
works of the Britons, nor have they ever pretended to any more
than to the repairs of them. Fleet, Lond. Scot. the spring
or flowing water place side. Flegg, Norf. the inclosing water place confines. Flintshire, or Teyengle, the inclosing
water

water place or Englifh fide, fhire, and accidentally the fhire of flintftones. Floddon, N-hum. the fide fpring water or flood place town. Flora, the goddefs of flowers, on the furrounding fpring water place or flowers. Florius, a river of Spain, the furrounding fpring water place Fuel, Som. the flowing fpring water place. Hence Flaby, by the fpring water place; Flag, the fpring water place confines or fea fide, where the fpring ceafes, and the firft ftones were found; Flame, the furrounding fpring of light; Flank, the inclofing fpring water place fide; Flannel, the inclofing or covering fpring of wool; Flap, the flowing fpring water place part; Flafhy, the fide fpring water place; Flafk, the fide fpring water place fhut; Flat, the fpring water place fide; Flatter, on the fpring water place fide; Flatulent, the fpring water place fide bank; Flavour, the fpring water place border; Flaw and Flow, the inclofing fpring water place; Flax, the covering fpring part; Flay, the fpring place of growth, Flea, the flowing water place or race; Fleet, at the fpring water place; Flefh, below the fpring place; Flexure, the border fpring water place; Fling, the inclofing fpring water place; Flirt, on the fpring water place or flowing fide; Flitch, the inclofing fide place; Float, at the inclofing fpring water place; Flock, gathering or inclofing fpring water place fide, or its lock part; Flook, a fhut or hook on the flood; Floor, on the inclofing fpring water place; Flounce, on the inclofing fpring water place or flowing fide; Flounder, flowing under or below the water; Flue, the flowing place of fmoke or other fluid; Fluid, the flowing divifions; Flufh, below the fpring water place; Flufter, on the fide of the fpring water place; Flute, the fide fpring water place; Flux, the fide fpring water place confines or action; Fly, the high or flowing fpring.

FOLEY and Foliambs, Worc. Der. the circle or furrounding fpring place. Folkingham, Linc. a vale on the furrounding water place confines. Folkftone, Kent, on the furrounding water place fide or border, or a town thereon. Ford, Devon, N-hum. the border fpring water fide or way, or a ford. Fordic or Fordwic, Kent, the upper way, ftreet or ford. Forden, Scot on the border water fide, or a vale thereon. Foreft, as that of Windfor, the furrounding fpring water part lower fide; or below the fame, or places ufually left uninhabited on the borders of countries, to prevent any fudden fkirmifhes that muft have otherwife happened. Forfar, or Ouil. Scot the border or fpring water circle part, or a high way in a fecondary fenfe. Formby, Lanc the furrounding fpring water part or habitation. Fornefs, Lanc the furrounding fpring water part lower

M   has

fide or promontory. Fofs dike or ditch, one of the four principal high ways leading from Lincoln through Caftle, by Newwark in Nottinghamfhire, Leicefter, Sharnford, Stretton on Dunfmore, and Stretton Super-ffofs in Warwickfhire, Lemington, Stow on the Would on the confines of Wilts and Glocefterfhire, to Cirencefter, upon which the Coritani extended their limits weftward fignifies. the border or furrounding water part fide, dike, or ditch. Fotheringhay, N-ham. the high inclofing fpring water part, or the fodering mead. Fouldon, Scot. on the border fpring water place, or a town thereon. Foulis, Scot. the border fpring water place fide. Foulnefs; Yorkf. the border fpring water place promontory. Fourmanteen, Scot. the furrounding fpring water divifion part. Fowy or Foy, Corn. the border fpring part. Foymore, the border fpring part more. Hence Foam, at the circle fpring part; Fodder, on the circle fpring part fide; Foe, the circle fpring or border part; Fog, the circle fpring part confines; Foible, the circle fpring from flowing part; Foil, the circle fpring from flowing; Foift, the fide below the circle fpring part; Fold, the circle fpring water place fide; Foliage, the circle fpring water place edge; Folk, the circle fpring water place confines; Folkmote, the circle fpring water place confines meeting, or a general meeting of the folks or inhabitants thereof; Folkland, the circle fpring water place or common people's land or copy hold; Follow, the circle fpring water place from the fpring place or low; Folly, on the circle fpring water place; Foment, at the circle fpring edge; Fond, on the circle fpring fide; Font, the circle fpring head or upper part; Food, at the circle fpring part fide; Foot, the border fpring part; For, the circle furrounding fpring part; Forage, the furrounding fpring part edge; Forbear, the fpring part from being upon; Forbid. from the fpring water part fide; Force, the furrounding fpring water fide or lower part, or at its entrance into the fea; Ford, the furrounding fpring water part fide or way; Fore, the furrounding fpring water part, which is the foremoft part; Foreign, on the border water confines; Foremoft, the furrounding water loweft fide border; Foreft, below the border fide, as the Hercynian below the German borders, extending from Alfatia and Switzerland to Tranfylvania; Forfeit, the border part, where fines and duties were payable; Forlorn, before the inclofing water place; Form, the furrounding fpring water parts; Formerly, the fea furrounding water part place, as it was our former refidence, or as it first furrounded the whole earth; Forfake, below the furrounding fpring water part confines; Fort, the furrounding fpring water part fide or inclofure;

Fortune,

Fortune, on the border fide; Forty, four ties in, fides, borders
or tens; Forward, towards the border ; Foſſe, the circle ſpring
fide ; Foſter, the circle ſpring fide water part; Foul and
Fowl, the circle ſpring water place ; Found, on the circle
ſpring fide ; Four, the circle or ſurrounding ſpring water
part of four fides, or the borders ; Fourth, the four fides or
borders

FRAMLINGHAM, Suff. a vale on the ſurrounding water part
place. Frankfort on the Maine, the incloſing ſpring water
part confines fo t; Frankley, Worc. the incloſing ſpring water part confines place, or dwelling place, the Franks being
in poſſeſſion of this part in the time of the Romans. Franks,
Germany, the incloſing water part lower confines, or a free
country or nation, as being ſituated on the ſea and Rhine lower
fide or ceaſt , and conſiſting of the Friſii, Sciamb i, Salii, and
Amſivari, of that ſignification, and of the Chamavi, Bructeri,
Chauci, and Catti, ſo diſtinguiſhed by their being keepers of
the French borders, but they ſeem to be a mixture of Cumbri, Gauls, and Phrygians. Fraumouth, Dorſet, the ſpring
water part or river mouth. Freſton, Linc. the lower or fide
water part town. Friburn, Kent. the part at the fide or below the incloſing ſpring part. Friſmerk, Yorkſ. the part below the ſurrounding water or ſea confines, or the meres.
Frith, Wales, the free or undivided part, or liberty. Frode-
ſham, Cheſ. the border water lower fide vale or dwelling place.
Frome and Fromcton, Wilts, Som. the ſurrounding ſpring
water part or town. Froſwell, Eſ. the ſurrounding ſpring
water part fide ſpring or dwelling place. Hence Fraction,
the fide or confines water parts, which had broke or divided
from each other; Frail, the water part place; Frame, the
ſurrounding water part, as waters divide the earth into parts
ſo as to form countries; Frank, on the incloſing or ſpring water part confines; Fraternal, on the ſame incloſing ſpring
water part fide; Fraud, dividing or leſſening the border or
ſpring water part fide; Fraught, at the incloſing ſpring water
part confines; Freak, the ſpring water part confines or border ;
Free, the ſpring water part; Freedom, the ſpring water parts,
limits, or dominion, which was that of the tribes or bands of
Gomer, whoſe ſupreme power was in the witen gemot and
foikmot, and from whoſe junction our parliaments; Freehold,
the free or flowing poſſeſſions; Freeze, the ſpring water part
lowering or ſtanding; Frequent, the incloſing ſpring water part
going; Freſh, the leſſer ſpring water part; Fret, the fide
ſpring water part; Friar, on the ſpring water part; Friday,
the ſpring water part fide diviſion; Fridburgh or Frank pledge,

free

free on the inclofing water place confines or edge, or the fecurity given by the decena or tything men, to anfwer for each other for their forthcoming at the view of frank-pledge, on their obtaining the liberty of going beyond the boundaries; Friend, on the foe fpring water part; Frijid, the fpring water part together or fhut; Fringe, the inclofing part; Frifk, the fpring water part at the inclofing water fide; Frith, the fide fpring water part or peffage; Frizle, the fpring water part fide place; Fro and From, the furrounding fpring water part or the border; Frock, the furrounding part covering; Frog, furrounding the water parts; Front, on the border fide; Frott, the fpring water part at lowering; Froth, the furrounding water fide; Frugal, inclofing or fhutting up the fpring water part; Fruit, the fide or furface fpring parts; Fry, the fpring water parts.

FULBOURN, Camb. the fpring water place neighbourhood. Fulham or Fullan, Midd. the fpring water place bank or vale. Furle, Suff. the fpring water part or dwelling place. Furey, Ire. the fpring water part. Furfham, Yorkf. the fpring water part fide vale. Furii, the fpring water parts. Furnival, Yorkf. the inclofing fpring water part vale. Hence Full, the fpring water part place; Fumble, about the fpring water part place; Fume, the furrounding fpring part; Fun, the fpring part in; Fund, the fpring part infide; Funeral, on the fpring water part; Funk, on the fpring part confines; Funel, in the fpring water place; Fur on the fpring part; Furlong, the fpring water place bank or length; Furnace, the fpring part in the fide; Further, the fpring part fide water; Fury, the fpring water or fire part; Fuffe, the fide fpring, Fufy, at the fpring fide; Future the fpring part to the fide.

GABRANTOVICORUM Portus or Sinus, Yorkf. the port, bay, wick, or ftreet of the water part coafters, Brigantes, or Kentifh Briges or Britons, or of the Cymbrian fea. Gabrocentum or Gatefhead, Dur. the Brigantes lower fide or gate. Gadeni or Ladeni of Teifidale, Tweedale, Merch, and Lothian, or Lodencian, feated between the mouth of the Tweed and the Firth of Forth, or Edenburrow, Scot. the inclofing water fide or vale, or the country below the water fide Gades, Spain, the lower water fide or confines. Gaer, the inflection of Caer and Gaervechan, N-Wales, an inclofed place or city, and a city on the fpring water and a little city. Gael, Ar Gael, or Argyle, the fpring water place, vale or Gaulifh country poffeffion or borders. Gaffelford, o Camelford, Corn. the furrounding fpring water place way, ftreet or ford;

an l

and gabel, gaffel, or gavel in a fecondary fenfe, fignifies fuch
a holding or its rent. Gainefburrow, Linc. the inclofing water fide burrow. Galanthis, Alcmenas waiting maid, the inclofing water place fide, bank, or haven. Gallana or Walwic,
W-mor. N-hum. the inclofing fpring water place bank, ftreet,
or fortification. Gallovidia, Galloway, or Galwallia, Scot.
Ire. the Gaulifh or Welfh way, ftreets or vales, on Watling-
ftreet, from North Wales to Dunbarton in Scotland, and in
Maio, Longford, and Gallway in Ireland, a proof of the Welfh
and Englifh, and the other Irifh and Britons being of the fame
origin. Gallena or Wallingford, Berks, the inclofing water
place, way, ford, or Watling-ftreet. Galloglaffes, Ire. the
Gaulifh inclofing water place guardians. Gallow-dale, Linc.
the inclofing water place, or Gaulifh vale. Galtres, or Cale-
terium Nemus, Yorkf. the inclofing fpring water place, a
Gaulifh, Celtic, or Caledonian grove or foreft; Gale being
the inflection of the Britifh Kelley, or Calli, the origin of
the Greek Keletes, the Roman Galli, the Irifh or Scots
Gael, and the Englifh Celtes; and thence vale and
wales in this refpect, though vale in a more primary fenfe
comes from bale and fale. Gallicia, Lufitania, or Portugal,
the inclofing fpring water place, or Gaulifh fides, confines or
port. Galtrim, Ire. the inclofing fpring water place or Gaulifh rim or Trim borders. Gallia Togata or Lombardy, Brac-
cata or Narbon, Provence, and Dauphiny, and Commata or
Belgica, Celtica, and Aquitanica, the inclofing fpring water
place or vale on the borders, upper fpring water parts and commots. Gallacum or Whelp caftle, W-mor. the inclofing fpring
water place part or vale. Galli, Carberi, or Corybantes, the
priefts of the Phrygian Cybele, on or the guardians of the inclofing fpring water places or borders, Cybele being the fpring
water place inclofure or their chief. Hence Gable, the inclofing or border water place, which floped, and was a gab-
bling place from the great concourfe of different people;
Gabion, on the inclofing water part; Gad, the inclofing water fide, where fairs and publick meetings were held; Gaffer,
an holder; Gaffles, holds; Gage, the inclofing water edge,
which confined, pledged, or place edged; Gain, the inclofing water or border, which muft be a gain, or a gain at the
market held thereat; Gale, the water place fpring; Galiot,
at the inclofing water place fide; Gall, the inclofed fpring or
powerful water place; Gallant, the inclofing water place bank
fide; Galley, on the inclofing water place.

GAM, the inflection of Cam, Wales, the furrounding water.
Gamlinghay, Camb. the cam or furrounding water high bank.
Gangani, Concani, or Conaght, Ire. the inclofing fpring water

confines, probably the descendants of the British Cangi, Casigani, Kentishmen or Saxons. Ganoe or Conwy, N-Wales, the inclosing spring water on the sea confines. Gannodurum, the inclosing spring water part Garienis, Suff. the inclosing water on the lower side. Garienonum or Yarmouth, the Yar or Garien mouth. Garioch, Scot. the upper borders. Garnsey, an island, in the inclosing water island. Gargrave, the inclosing or border water warden. Garth and Garthum, Wales, Yorkf. the inclosing water side or an inclosure. Gafwood, Norf. the lower inclosing spring side or wood. Gatefhead, Gabrocentum, or Pandion, the inclosing water side head, gate, gatehouse or townfend. Gatton, Surry, the inclosing water side gate or town. Gavel-Gabel or Gafel-kind, Kent, Wales, a joint inclosing spring water place or a holding in kind. Gaunlefs, Dur. the lower or leffer inclosing spring water. Gaufenna, on the lower inclosing spring water. Gawthorp, Yorkf. the inclosing spring water gate. Hence Gambler, on the surrounding water or a crooked place or player; Game, the surrounding water confines; Gammon, the surrounding crooked part; Gander, on the inclosing water part side; Gang and Gangway, the inclosing water confines and way; Gantlet, the inclosing water side let, leet or hindrance; Gaol, an inclosed place or shut; Gap, the part or division of an inclosure; Grab, the covering part; Garden and Garth, on the inclosing water side or inclosed parts or sides; Garland, the inclosing water bank side or the land borders, borderers or gargraves; Garment, Garner, Garret, Garrifon, and Garter, covering and inclosing parts; Garthman, the wear man; Gash, the side inclosing; Gafp, the side part shut; Gate, the side inclosing; Gather, on the side inclosing; Gaudy, the inclosing side; Gauge, the inclosing spring water confines; Gavelman, an holder, underholder or tenant; Gaufe, the leffer covering; Gay, the inclosing water; Gaze, below the inclosing water side.

GEDDINGTON, N-ham. on the inclosing water side town. Gedney Moor, the inclosing water side moor. Geminham, Norf. on the surrounding water confines vale, where the foccages, surrounding water edge or Saxon tenure, in paying with day's labour inftead of money, was kept up according to Spelman. Geneu, Corn. Wales, the inclosing spring water mouth. Genevil, Ire. the inclosing spring water mouth dwelling place or village. Genevifh, the inclosing spring mouth side or dwelling place. Genilla, Gloc. the lower or side inclosing water or the Severn Genoua, an empory of Italy, the inclosing spring mouth. Genunia, Gueneth, Venedotia, Ordovices, or North Wales, the inclosing spring water sides, streets,

ſtreets, vales or borders or the Veneti or wenta incloſing borders; Gerion of Spain and Cerberus, the weſtern water borders, or Italy, Spain, and Gaul, which were in the poſſeſſion of
the Germans or Ger-mons, and Greeks and Phrygians. and
from whence Hercules, as from hell or the undrained lower
country, brought the three headed dog Cerberus, to a place
where he ſaw the light, and vomited or diſcharged its waters
into the ſea, and this three headed nation ſeems to be the
foundation of the Gauliſh, German and Celtic. German,
Corn. the incloſing or border water part or haven. Gerno,
Norf. the incloſing water. Gerard, on the incloſing water
ſide. Gervii, the incloſing ſpring water dwellings. The ancient Germans ſeem to be comprehended in the Allemanni,
the alley, galley, or Gauliſh parts or men; Boii, the Po or
border ſpring water; Bigians, Friſians, Pruſſians, or Britons, on the water or ſpring water parts or ſides; Cymberi,
Marcommani, or Gomerimanni, Veneti, Vandales, Vindili,
or Werni, dwellers in vales, or on the river ſides; Danes, on
the water part; Germans, Ger-mæons, or Hermeones, Her-
manduri or Germanduri and Suevi, the ſea borderers and
ſpring ſide dwellings; Gothones, Gepidæ, and Dacians, the
incloſing water ſide or the Danube; Hunns and Burghundi,
on the high incloſing ſpring water ſide or confines part, or the
dwelling parts on the Danube; Helveti, the hill dwellings or
poſſeſſions; Langobards, the ſea bank ſides or countries;
Saxons, the lower incloſing water or ſea confines; Teutones
or Deutons, on the border ſpring ſide dwellings; and Volcæ
Tectoſages or Belgæ, the upper border vale confines. Thoſe
may be reduced to the Cumberi, or Gauliſh or Britiſh race,
the Dacians, Friſians, Germans, and Teutones; and theſe
again to the Dacians. Cumberi, and ſea Germans, the reſt of
the German names being mere local appellatives. As to the
original of the Germans, ſome bring them from Aſkenas, the
ſon of Gomer, which as to the Cymbri, Friſians, and ſome
other branches ſeems to be true, but the old ſea Germans were
a mixture of the Lacedemonians, and other Greeks and To-
germans, and the preſent Germany contains ſome of every people
of Gaul and the Britiſh iſlands; as do thoſe of the German
people; though, to gratify vanity, and the Roman policy of
promoting diviſions, great pains have been taken to darken
the genuine origin of thoſe people. Hence Gear. a covering
upon; Geld or Yield, the incloſing ſpring water place ſides,
which yields a rent or tribute as well as growth; Gelly, an
incloſed or ſhut water place; Gender, on the incloſing water
ſide; Genealogy, a diſcourſe on the incloſing water confines

or race; General, on the inclofing water place; Generate, at the inclofing water; Genna or Gehenna, the mouth, fink, fwallow, or hell; Gentility, on the inclofing water fide part or property; Genuine, within the inclofing fpring; Geography and Geometry, the earth or furrounding parts writing and meafuring; German, the inclofing or border water part; Gerinmate, at the border edge; Get, inclofing the fides.

GIDDY Hall, Eff. a hall at the inclofing water fide. Gilefland or Geltflonde, Cumb. the inclofing water place below the bank fide or land. Gilling, Yorkf. the inclofing water place confines or dwellings. Gillingham, Kent, the inclofing water place vale or foreft. Gilfhaughlen, W-mor. the inclofing fpring water place bank. Gilpen, Cumb. the inclofing fpring water place end or head. Gingrave, the furrounding water or border warden. Ging, Ing, and Ingerfton, Eff. the inclofing water confines or town. Gingrave, the inclofing water warden. Menas-Inge, the inclofing water lower part. Marget ing, the mere gate confines. Gipfeys or Vipfeys, Yorkf. the fpring rifing up from below. Girwy or Jarrow, Dur. the inclofing fpring water, border or dwelling place. Girnego, Scot. on the fea border. Gilburgh, Yorkf. the inclofing water fide fpring part confines or burrow. Gues, Gloc. the lower or fide fpring water confines. Hence Giant, on the inclofing water or border fide, or the guardians of the borders, like the giants, Abarim, Avim, Captorim, Philiftin, and Rimmon, who were guardians of the rims or borders in the land of Canaan; Gibberifh, on the border fide; Giddy, at the inclofing water part, or fwimming in the head; Git, the inclofing fpring water part to or at the border; Gild, inclofing the fides; Gim, the covering; Gimlet, to the water place covering; Gin, inclofing; Gingle, the inclofing water place confines; Gipfey, below the fide water part; Gird and Girdle, furrounding the fides; Girle, the inclofing fpring water place; Gus, Goofe, or Geefe, the inclofing fpring fides.

GLADMOUTH, Kent, the divifion fpring water place mouth. Glamorgan or Glanmorgan, S-Wales, the fea inclofing water place confines, bank fide, or country. Glamys, Scot. the inclofing water place fide or country. Glan or Clancatting, Scot. the border water keeping clans, like the Catti on the Rhine. Glancolkin, Irc. the inclofing water place confines bank. Glandilagh, S-Wales, Irc. the divifion or boundary water place bank, land or country. Glandford bridge, Linc. the inclofing water bank way or ford bridge. Glanjores, Irc. the inclofing water place fhore. Glan lyn, Wales, the lake bank.

bank. Glanaventa, Glenwelt, or Caervoran, N-hum. the inclofing fpring water or Veneti parts or banks. Glafcow, Glaf-
ton, Glaffen, Gafnith, Glafienbury, Avalon, and Petra ad
Glottam, Corn. Som. Lanc. Scot. Ire. the inclofing fpring
water place fide, bank, vale, burrow or town, and in a fecondary fenfe, a green place at the river fides, famous for faints
and religion. Gledmore, Midd. the inclofing water place
fide moore. Glan, Glen, or Glin, Norf. Linc. N-hum;
Scot. Wales, Ire. the inclofing fpring water place fide, bank
or vale, or the lans, lands, lancs, alleys or ftreets of the Gaulifh migrations. Glencarn, Scot. on the inclofing water border. Glendale, N-hum. on the inclofing water border vale.
Glendour, N-Wales, the divifion water part bank or vale.
Glocefter, Glevum, Clevum, Gleancefter, or Gloui, the inclofing fpring water place, lock, key, or city. Glotta or Cluid,
Scot. N-Wales, the inclofing fpring water place fide or coaft.
Hence Glacis, the inclofing water place fide or flope; Glad,
the inclofing water p'ace fide; Glance, the fide of the fight;
Gland, the fide divifion; Glanders, the gland water; Glare
or Glair, an inclofed light; Glafs, the fpring water place fide
or green like grafs; Gleam, the furrounding light; Glean,
to clean; Glebe, the inclofing fpring part; Gleet, the inclofing water place fide; Glib, the flowing water or wet part;
Glide, the inclofing water fide; Glimmer, a dead furrounding light; Glimpfe, the furrounding light part low; Glitter,
the light upon or ftriking upon a part or water part; Globe,
the furrounding place or water place or circle part; Gloom,
the furrounding light fhut; Glory, on the circle water part
or country, where glory may be acquired; Glofs, the inclofing
water place fide, confines, key or fight; Glove, a covering
part; Glout, the light fhut out; Glow, the furrounding light
fpring; Glue, the inclofing fpring; Glut, the inclofing fpring
fide.

GOBANIUM, or Abergaveny, S-Wales, the inclofing or
border fpring water parts. Goccium, by the inflection of
Coccium, Lanc. the inclofing water confines or city. Go-
dalmin, Surry, the inclofing water edge vale. Goderick,
Heref. the upper border or inclofing water part. Godman-
ham, Yorkf. the inclofing water fide part vale. Godoleen or
Godolphin, Corn. the fea vale confines or edge. Godfhill,
Hants, the inclofing water fide hill. Godftow, Oxon. the inclofing water fide ftation. Godwick, Norf. the inclofing water fide wick, ftreet or fortification. Gogarth, N-Wales, the
fea inclofing water part or promontory. Gogmagog hills,
Camb. the furrounding water, Magog or great hills. God-

N

manchester or Durioaponte, Hunt the Pontes or Wapoonta-
ches tide water part. Gonshill, Suff. the inclosing water side
hill. Godman's fields, Midd. the fields or dwelling place at
the inclosing water tide. Goodwin sands, Kent, the inclosing
spring water at the sea side sands. Golden, Corn. Heref. the
inclosing water place side or vale, or the golden vale. Gomer
by the inflexion of Comer, the sea or side surrounding water
confines. Gorambery, Herts, the surrounding spring water
part border. Gorleston, Suff. the inclosing water place side
town. Gosfcford, N-hum. the inclosing spring water side
ford or way. Gothia or Gothland, an island in the Baltic,
the water surrounding parts or possessions, or sea coasters descended from the Dacians and the Greeks of the golden fleece.
Govenny, S-Wales, the inclosing spring water parts. Gowle,
Yorks. the inclosing spring water place. Gowry or Gower,
Scot. N-Wales, on the inclosing spring water. Hence Go,
the circle or inclosing water and goat, which goes; Goad,
the goat; Goal, the go place; Gob, the go part, Gobble, on
the go part; God or Doge, the governor of the world; Gold,
the sun division; Good, the inclosing water side; Gore, the
going or running water or a shore; Gorge, the going water
confines; Gormandize, the going water part lower side;
Govern, on the inclosing spring water confines or country;
Gown, a covering upon.

Gradivus or Mars, on the surrounding water or sea side.
Grampian or Grantham, Scot. the high border parts. Gram-
pound or Grandpont, Corn. on the inclosing spring part side
border. Greaton, N-hum. the border part or warden town.
Grant, Camb. on the side water part or border. Grantham,
Linc. the grant vale. Granthorpe, Cumb. the side inclosing
water gots. Gravelend, Kent, the inclosing spring water part
or warden's end. Grenvil-chester-bith-holme-wich-haugh-
law-sted, Corn. Som. N-hum. Kent, W-mor. Lanc. Scot.
Suff. on the border or inclosing water village or dwelling
place, city, passage part, island, wick, street or fortification,
high spring water confines, spring water place and station.
Gretham, Dur. the border water tide vale. Gretland, Yorks.
the border water side bank side, land or country. Grey,
Cumb. the inclosing water confines or border, or its warden.
Griffin or Griffith, the inclosing water part, or border side or
edge, or its warden. Grimsby, Linc. the surrounding border
isle part or habitation. Grims-car-ton-hill, Yorks. the border
water side city, town or hill. Gromlech, N-Wales, the surrounding border water place confines or rocks. Gron, Camb.
the water inclosing borders, dams or fens. Grooby, Leic.

the border part or habitation. Hence Grace, the inclosing water confines or border side; Gradation, on the inclosing water sides or the degrees; Graduate, at the inclosing water or border side; Graft, the border part side; Grain, in the inclosing water; Grammar, on the surrounding border parts; Grand, on the inclosing water or border side; Grange, the inclosing water confines or edge; Grant, on the border side; Grapple, the inclosing water part place; Graip, the inclosing water lower part; Grass, the lower border side or surface; Grate, the inclosing water sides; Gratis, below the inclosing water sides or the commons; Grave, the border water part, which were the usual burying places; Gravel, on the inclosing water place; Gray, the border water; Grease, inclosing the sides; Great, at the inclosing water confines or border; Grecian, on the inclosing water, or water inclosed confines or islanders; Greedy, the inclosing water side or division; Green, on the inclosing water or the grass; Greet, inclosing the sides, Grim, the border man or giant; Grimace, his action; Grin, the inclosing water part in; Grind, on the inclosing water part side; Gripe, inclosing part; Gristly, the inclosing water side place; Groat, at the border; Grocer, the border side man; Groin, inclosing the border; Gross, the border side; Grot, a surrounding part; Grove, a covered part; Ground, on the border side; Group, the border or sand part; Grow, the border or ground spring; Growth, the ground spring parts or divisions; Grub, the ground springing up; Grudge, the inclosing water edge; Grumble, the surrounding water place.

Guidi or Guith, Scot. Wales, Jiants, the side or division spring water. Guenta, Venta, or Wenta, or rather Kenta Belgarum, Iceenorum, and Silurum, the inclosing spring water confines, or the first Veneti possessions of the Belgæ, Iceni, and Silures. Guilford, Surry, the inclosing spring water place, ford or way. Guis or Gwis, Wales, the side spring water. Guiscard, the side spring water guard. Gunrock, Scot. the surrounding spring water vale. Gurmon, the border or man on the surrounding water or sea, or a German. Guithlac, the side or separation spring water place confines. Guti, Goti, Gætæ, Jutæ, or Goths, one of the people who came here from Germany, the sea or river sides. Guy-cliff, War. the inclosing spring cliff. Gwydrin, Wales, on the side spring water part. Gwyneth, Venedotia, or N-Wales, the inclosing spring sides, vale, or Veneti boundary or borders. Gwithil Pictiad, or Calledonia, on the inclosing spring sides, upper parts or hills. Hence Guard, on the inclosing spring side, Gudgeon, the spring edge one; Guess, below the spring;

Guest,

Gueſt, below the incloſing ſpring ſide, or under one's roof or ſhelter; Guide, the incloſing ſpring ſide; Guild and Guilt, the incloſing ſpring water place ſide or yielding; Guiſe, below the covering; Gulf, an incloſing or incloſed water part; Gull, on the water place confines; Gum, the ſurrounding ſpring confines; Gun, the incloſing ſpring; Guſh, the ſoring water ſide; Guſt, the leſſer incloſing ſpring diviſion or ſide; Gut, the incloſing parts; Gutter, the incloſing water part; Guttle, on the gut.

Habitancum or Riſingham, N-hum, an abode or ſtation at the incloſing ſpring water place or vale ſide. Hache, Wilts, the incloſing ſpring water or a cut. Hackington, Kent, a town on the incloſing water confines. Hacomb, Devon, the incloſing water vale. Hackney, Midd. the incloſing water confines, the uſual place of traffick. Haddon-ley-ington, Der. Suff. Wilts, Oxon. N-ham. Scot. the incloſing ſpring water ſide town, place or palace. Haia, Hagæ, Haigh or Chaia, Surry, Lanc. W-mor. the incloſed water parts or other incloſures. Haile, Corn. the high or incloſing flood ſalt water place. Hailweſton, Hunt. the incloſing ſalt ſpring water ſide town. Hairholme, or Hairum for Chairum, the iſland or water ſurrounding part. Hale and Hales, Gloc. the high or incloſing ſpring ſalt water part. Haleſworth, Suff. its ſide ſpring or dwelling place. Haliburton, Scot. the ſalt ſpring water part town. Halifax, Yorkſ. the hilly ſpring water confines part. Haling, Kent, the high or incloſing ſpring water place. Hall, Corn. on the high water place. Halton and Halyſton, N-hum. the high water place ſide town. Ham for Cham, the inflection of Cam, and Hamden for Chamden, Bucks, the ſurrounding water confines vale, and when vales became the uſual places of reſidence, a home in a ſecondary ſenſe. Hampſted, Midd. Herts, the vale or home part ſide or ſtation. Hamble or Homlea, Hants, the ſurrounding ſpring water place vale or home. Hameldon, Dorſet, the ſurrounding ſpring water place ſide or hill. Hampton or Hanton, Midd. S-ham. Heref. the ſurrounding or incloſing ſpring water part town. Hanford-bridge-mere-over-well-worth, Devon, Salop, Cheſ. N-Wales, Germany, Oxon. Midd. the high ancient or incloſing ſpring water place, way, or ford, bridge, ferry, ſpring place, and ſide ſpring or dwelling. Hans, the lower incloſing ſpring or ancient confines. Harbottle-bury-ham-dwicke-field-ville-ingworth-lech-lſey-nham-petre-row-tiepole-wichwood, N-hum. Leic. N-Wales, Kent, Yorkſ. Der. Gloc. Midd. N-ham. Wilts, Som. Dur. Eſſ. the high water or incloſing clofing fpring water fide dwelling place, burrow, fide vale, fide ftreet or fortification, dwelling place, village, inclofing fpring fide, upper place confines or rocks, place fide, on the vale, fide part or rocks, fpring border, place fide poole, ftreet or fortification, and fide or wood. But H became a radical by dropping the C, from Ch the inflectory of the radical C, which has been alfo varied into Sh, whereby a new fort of dialect has been framed from the Celtic, as from Car or Char into Har and Share. Haflinbury, Eff. the lower inclofing fpring water place dwelling or burrow. Haflings, Suff. on the lower water confines. Haflewood, the lower or leffer fpring water or wood place. Hatfield, Yorkf. Herts, Eff. the high or inclofing fpring fide dwelling place. Hatherel, S-Wales, the hill inclofing water place. Hather, Linc. Chef. the high fide fpring water part. Havant, Hants, the inclofing fpring water or haven part. Haverford or Milford, S-Wales, the inclofing fpring, haven, or furrounding water place, ford or way. Havering, Eff. the high or inclofing fpring water confines. Haw, Devon, the inclofing fpring. Hawghley, Suff. the inclofing fpring water confines place. Hawick, Scot. the inclofing fpring water confines, ftreet or fortification. Hawfted, Eff. the inclofing fpring ftation. Hay for Chay, Cay or Cae, the high or inclofing fpring confines. Hayward, warden of the inclofures. Hence Habit and Habitation, on the inclofing fpring water part fide; Hack, Hackney, Hag, and Haggle, the inclofing water place confines, or a cut; Hail, the inclofed or fhut flowing water or rain; Hair, the covering fprings; Halbert, the high or inclofing fpring water part fide; Halcyon, on the water place confines; Hale and Hall, on the high fpring water place; Half, the inclofing fpring water place part; Halloo, on the high fpring of calling; Halo or Hailo, the fun's circle; Halt, the inclofing fpring water place fide; Halter, on its fide; Ham, Cham or Thigh, furrounding the water fides or confines; Hamlet, the furrounding fpring water place fide, houfe, poffeffions or leet; Hammer, the great ham, from its refemblance to a man's thigh, leg and foot; Hamper, the furrounding part; Hanaper, the inclofing or fhut up part; Hand, the high or fpring inclofing part or end; Hanker, on the inclofing fpring water confines; Hap and Happy, the high or inclofing fpring part; Harangue, on the inclofing fpring confines; Harafs, the inclofing fpring lower fide; Harbinger, the inclofing fpring end man; Harbor, the inclofing fpring fea or border part; Hard, the high water fide, which is commonly rocky; Harm for Charm, and Harmony, on the furrounding water confines; Harnefs, inclofing or covering on the

the lower place; Harp, the high fpring part; Harrow, on the
fprung up ground, Plow being a part fprung up; Hart, the
inclofing fpring water part; Harvell or Charwell, the fpring
below or under covering; Hafh, fmall parts; Hafp, the leffer
fhut part; Hat or Chat, the upper covering; Hatch, at the
covering; Haven, the inclofing fpring water confines; Have,
the inclofing fpring part; Haughty, the inclofing fpring water upper fide or poffeffion; Haul, the high or fpring water
place; Haum, the furrounding fprings; Haunch, the inclofing fpring water place confines; Haunt, the harbour, as a
place of refort; Havock, a high fea fpring; Haw, the high
fpring; Hawk, the high confines fpring; Hay, the high fprings;
Haze and Hazard, for Chace and Chace hard, on the inclofing
water or border lower fide, or on the chace.

HEA, the high water or fea. Headon fenfeld-ly-rthy-th-vy,
Yorkf. N-hum. Staff. W-mor. Devon, the high water or fea
town, field or dwelling place part, place, on the fide, fide or
heath or dwelling. Hebrides or Hebudes iflands, W-Scot.
the high water part lower or weftern fides, parts or iflands.
Heckham, Norf. the high water confines vale. Heddington
for Chedington, Wilts, Oxon. the high or inclofing fpring
water fide town. Heil and Heillion, Corn. Dor. the high or
falt water place fide town. Heitelbury, Wilts, the high water
lower fide fpring dwelling part or burrow. Helaugh, Yorkf.
the high water place confines. Helbeck, Yorkf. Cumb. the
falt water place brooke. Hellens, the falt water place bank.
Hellan, Scot. the falt water place bank. Heming iord and
ton, Suff. Yorkf. the inclofing water edge, ford, way or town.
Heneti, Cheneti, Gueneti, Veneti, Venedi, Wenta and Venedotia, Italy, Gaul, Germany, and the Britifh iflands, the
high, ancient or inclofing fpring water fide or vale poffeffions
or poffeffors. Henbury-don ham-ley-ton, Gloc. Midd. Norf.
Worc. War. Oxon. the high ancient or inclofing fpring burrow, town, vale or home, or place, but the original radix is
Cen, which has been varied to Chen, Gen, Hen, Ken, and
Shen. Heortnefs or Heortu, for Cheortu, Dur. the high water border lower fide or promontory. Hep and Shap, for Chep,
W-mor. the high inclofing water part. Hepbury, on the high
water fpring or dwelling part. Heraclea or Cheraclea, Hercules pillars, Spain and Macedonia, the inclofing fp ing water
place lower borders. Hercules promontories, rocks, iflands,
ports, towers, altars, way, haven, and grove, in Mauritania,
Galatia, Britain, Carmania, Sicily, Sardinia, Italy, and
Germany, the inclofing fpring water place or vale lower borders or havens, and the migrations and coalitions of Greeks,

Germans,

Germans, Phrygians, Gauls, Celtes, and Cumberi along the same. For Hercules is said to be the son of Jupiter, a Titan, by Alcmena, of the Greek and German race, along the western spring water parts or havens; whereof Hercules was guardian; and to have married the daughter of Alesia, the lower water places or Celtes, and begat Galates, the founder of the Gauls, or in other words mixed with the Celtes, and founded the Gaulish nations. His cleansing the Augean stable, containing 30000 oxen, alludes to his opening, draining, dividing, and founding, or bounding by rivers, so many districts of the Cumbri or dwellers in vales His driving Gerion, the western water borders, or borderers, with his cattle out of Spain, alludes to his sending colonies from thence into Germany. His binding in 100 chains, and throwing out of hell, the three-headed dog Cerberus, seems to refer to his draining, dividing, and bounding the countries of the Titans, by rivers in Spain or Tartesa, then undrained, and deemed as the most western country where the sun appeared to set, to be hell by some of the ancients. These also seem to be the 100 headed and 50 bellied monster of Homer, named by the gods Briareus, Phrygians and Titans, and by men Ægeon, Greeks and Germans, who flung at Jupiter at once 100 rocks, sea banks, or promontories, which, on the retreat of the deluge, had formed along the sea coasts, preventing the discharge of the land water into the sea, which Hercules removed. And these Herculean colonies probably were the foundation of the story of the Trojan Brutus. Hercules Midacritus, the Hercules of the Mediteranean Crete, who in digging of the sea and river banks found tin ore. Hereford or Cherford, the border water way or ford. Herlaxton-tford-thy-tlebury-tland-ton-ty, Herts, Hunt. Worc. Devon, the border water place confines town, side ford or way, side or possessions, side place burrow, and side land, town or possessions. Hesperides, Chesperides, or Cassiterides, and as some suppose the Scilly islands, the lower water parts or border islands, but none in particular from whence the Greeks arbitrary name of tin. Hesperia, or Spain, on the lower including water part or border. Hessel, Yorks. below the border water place. Hexton, Midd. on the lower border part. Hethy or Chethy, one of the Orcades, the inclosing water side. Heveringham, Suff. the inclosing spring water confines vale. Hewet, the high or inclosing spring tide. Hexham, N-hum. the high or inclosing spring water confines vale. Heyford, Oxon. the high way on the inclosing spring water part. Akeman street, on the warring town. Hey or Chey, Lanc. the high or inclosing spring. Hence Head or Chead, the top or

covering

tovering part; Head borrow, the head of the spring water border; Heal, springing up; Heap, high part; Hear, the ear spring; Heart or Cheant, the inclosing spring water part; Heat, the high spring; Heath, the high water side; Heathen, the heath or water side men, or the Gothi Danes,; Heave, springing up; Heaven, the high water parts or clouds; Hedge, the high edge; Heed, the high side for the defence of the country, Height, the high side; Heir, the male border spring; Heiress, the female border spring; Hell, the low water place or the bottom of the sea, as heaven was in the clouds; Helm, over the sea; Help, the spring water place part; Hem, the surrounding spring or edge; Hemp, the surrounding spring or covering part; Hen, the ancient or inclosing spring or covering one; Hence, the inclosing or ancient spring side; Henceforth, from the side of the inclosing spring; Henceforward, from the inclosing spring side below; Heptarchy, the seven governments, or inclosing water confines or countries; Herald, on the border water side, or a challenger; Herb, the spring part; Herd, the spring side; Here, the spring on; Hereafter, after the spring upon; Heresy, below the bounds; Heriot, the circle side or border springs; Heritage, the heirs borders; Hermit, the surrounding spring water side; Hermitage, a hermit's shut or cell; Hero, the border water man; Herring; the inc'osing water border; Hesitate, at the inclosing spring or border side; Heterodox, from the inclosing spring border; Hew, an high spring, or the noise usually made at the time of hewing; Hey, an high spring.

HIERON, a promontory of Ireland, on the higher inclosing water or the longest, farthest or holy one. Hicrytha, Devon. the high or inclosing water or farthest border. Higham Ferrers, N-ham. the inclosing spring vale lower part. Highworth, Wilts, the high or inclosing spring water side. Hill, high on the spring water place, a bank, an high place or a hill. Hildersham, Camb. the high or inclosing water part side vale. Hilshaw, Yorks. the spring side high place. Hinchingbrooke, Hunt. a brooke on the inclosing water confines. Hinckley, Leic. the inclosing spring water place or palace. Hinderskel, Yorks. the inclosing spring water part cell, shut or castle. Hingeston, Camb. the inclosing spring water or ditch side town, as the borders of the east angles. Hirta or Hirth, Scot. the high water side or far part. Hithe, Hyth or Hyde Kent, Hants, Yorks. the side or passage part or port or the length of the side spring. Hence Hickup, springing up; Hide, the side covering; Hierarchy, the higher government; Hieroglific, the high sacred or inclosing spring water border sculpture;

Higgle,

Higgle, on the inclofing fpring water place ; High, the high or inclofing water confines ; Hind, the inclofing fpring, a ditch or ditcher ; Hinder, the ditch water part ; Hindmoft, the furrounding ditch lower fide ; Hinge, the inclofing fpring edge; Hint, the edge part ; Hip, the fpring upper part ; Hire, on the fpring or ftay ; His, the male or he fide, divifion or things ; Hiftory, on the border fides ; Hit, fpring at ; Hither, on the fide or fpring on the fide ; Hitherto, to the fpring on the fide ; Hive, the inclofing fpring or fwarm part: but all thefe radicals had originally c before the h.

Hockley, Bucks, the high furrounding fpring water confines place or a wet or miry place. Hodde, Dor. the inclofing fpring fide part. Hodni, S-Wales, on the inclofing fpring part. Hole Haven, Eff. the inclofing fpring water place haven. Holbourn, London, Scot. on the inclofing fpring water place or hilly parts or country. Holdernefle, Yorkf. the inclofing fpring water on the lower fide or promontory. Holefhot, Hants, the inclofing fpring water place fhut. Holgrave, the inclofing fpring water place warden. Holland Lane, Eff. Linc. Nott. the inclofing fpring water place bank fide or country. Helme, Cumb. Linc. Nott. Herts, the inclofing fpring water furrounded place or a river ifland. Ho'mefdale or Cholmefdale, Surry, its vale. Holt, Worc. Norf. N-Wales, the inclofing fpring water place fide or wood. Holy Head, N-Wales, the high or inclofing water place head. Hoo, Kent, the high or inclofing fpring part. Hooke, Ire the inclofing fpring water confines or fhut. Hope, N-Wales, the high inclofing fpring water part. Horefci or Horeftii for Corefti, Scot. the lower inclofing fpring water or Efk lower confines, or the lower borders, the Calledones being a general name of the highlands, though they were divided into Deu-calledones and Vecturiones. Horneby, Yorkf a part or habitation on the inclofing fpring water or border. Horne caftle, Linc. a caftle on the inclofing fpring water or border. Hornley, Yorkf. the inclofing fpring water or border place. Hortham, Suff. Norf. the border fide vale. Horton or Halifax, N-hum. Yorkf. the high or inclofing fpring water or border town. Horfted, Kent, the high or inclofing fpring water fide or border ftation. Houghton, Norf. the high or inclofing fpring water confines town. Hounflow, Midd. the inclofing fpring fide water place. Houth, Ire. the inclofing fpring water fide. Houburn, Scot. on the high inclofing fpring water part. Howden, Yorkf. the high inclofing fpring vale. Howgil, W-mor. the high inclofing fpring water or river fide. Howley, Yorkf. the high inclofing fpring water place. Hox and Hoxon, Suff. the high inclofing fpring water

O   confines,

confines. Hence Hoar and Hoary, the fhut fpring; Hoard, the fhut up fprings; Hobble, the fprings fhut up place; Hobgoblin, the inclofing fpring water confines bank; Heck and Hog the inclofing fpring water place, or wet or miry confines, and a hock of bacon; Hod, a fhut or clay part; Hoift, high or fpring up from below; Hold, the inclofing water place fides; Hole, the inclofing water place; Holy, the inclofing water place, as the place of religious worfhip; Home or Chome, the commot or furrounding water part fide; Homage or Chomage, an engagement to defend the borders and obferve the frank pledge, on being made free of the commot in refpect of poffeffions therein; Homicide, flaying a man without malice; Honefty, the inclofing fpring fide tye or bond; Honey, in the inclofing fpring; Honour, the whole inclofing fpring or border part, manor being but a part of the circle or border; Hood, the high fide covering; Hoof, the foot covering; Hook, an inclofed confines or edge; Hoop, a covering part; Hop, a fhut part or a part fhut; Hope, the high fpring part; Horizon, the covering on the lower fide; Horn, on the high inclofure, covering or top; Hofe, the lower covering; Hoft, at the fide covering; Hot, the high fide; Hovel, a fmall covering or cottage; Hound on the fpring part; Hour, the fpring circle; Houfe, the fide covering; How and However, what fpring; Howl, on the high fpring of calling; Hoy, the high inclofing fpring.

HUCKSTOW Foreft, Salop, the high inclofing fpring water ftation foreft. Hulebec, Salop, the high fpring water place brooke. Huis le Britein, the high fide fpring at the water part fide, or a houfe at the water part fide. Hull, Yorkf. on the high inclofing fpring water place. Humber or Chumber, Yorkf. the high furrounding fpring water part. Hume or Chume, Scot. the high furrounding fpring. Humbie or Chumbie, Scot. the high furrounding fpring part, vale or dwelling part. Huna, W-mer. the high inclofing fpring. Hungerborn, Wilts, the inclofing fpring water border neighbourhood. Hungerford, Berks, the inclofing fpring water ford. Hunftanton, Norf. a town below the inclofing fpring fide. Hunfdon, Herts, the inclofing fpring fide town. Huntcliff, Yorkf. the cliff at the inclofing fpring fide. Huntercomb, Heref. the high inclofing fpring water part vale. Huntingdon, a town on the inclofing fpring fide. Huntley, Scot. the high inclofing fpring fide place. Hurley, Ire. the high fpring water place. Hurft or Hurft, Hants, Suff. the high fpring fide covert or wood. Hurft caftle, the foreft caftle. Hufeley or Dorchefter, the inclofing fpring fide water place.

Hu or Hy, the high or inclosing spring. Hygre, the high water border. Hufcarles, Dorset, over the inclosing spring water place or border. Huftings, a house on the inclosing spring side for the determining disputes within particular limits. Hytching, Herts, at the inclosing spring water confines. Hence Hubub, a high spring part up; Huckleboue, the high spring place bone; Huckster, on the inclosing spring or border side; Huddle, at the inclosing spring water place; Hue, from the spring of light; Huff, springing off; Hug, inclosing spring; Hulk and Hull, the water inclosing place; Hum, the surrounding spring; Human, the surrounding spring part; Humble, the surrounding spring water place; Humid, the surrounding spring side; Humor, from the surrounding spring; Hump, the surrounding spring part; Hundred, the inclosing spring water part side divisions, or a centuria; Hunger, on the inclosing spring confines; Hunt, the inclosing spring side or forest; Hurdle, the inclosing spring water side place; Hurl, the high or inclosing spring water place; Hurly burly, the high spring water burrow place; Hurricane, the high spring set in; Hurry, on the spring; Hurt, the inclosing spring part or a cut; Husband, the inclosing spring side or house band; Hush, spring low; Husk, inclosing or covering the lower shut or core; Hut, at the inclosing spring; Hydra, the high or inclosing spring water part; Hydrometry, the high or inclosing water part measure; Hymen, on the high surrounding spring; Hyperborean, on the high spring water part border; Hypocrite, at the inclosing spring water part border.

Ial, N-Wales, the heights. Jarrow, Girwy, or Yeram, Dur. the surrounding or border spring water. Ibercan, or Iberi, Ire. Spain, on the inclosing water part. Icaldun, Ess. the inclosing water place bank or hill. Ivanhoe, Linc. the high inclosing water. Iſceni, Ikeni, or Iſceni of Suff. Norf. Camb. Hunt. and Iebene of War. the inclosing water, lower, island or Kentish confines or side, and as Tigeni, Simeni, and Iſceni Magni, the inclosing water, the sea or Kentish side. Ichborow and Iſciani, Norf the inclosing water confines or Iſceni burrow. Ichenild or Ikenild street, one of the four British streets, whose course was not only through the four Iſceni counties, but also through Staff. War. Worc. Gloc. the Iſceni street or inclosing water place side or high street. Ichlingham, Suff. the Iſceni inclosing water place vale. Ichford or Ickford, Bucks, the upper or Iſceni way or ford. Ickley, Dur. the Ikeni or inclosing water place. Ikentworth, Suff. the Ikeni or inclosing water side spring water. Ikenthorp, Suff. the Iſceni

water

water gate. Ikening, Suff. the Ikeni inclofing water or border, and from ing came ing-land, or an ifland in the Cymbrian dialect, as well as Ingles and Angles, an incorporation of the Ifceni with fome of the lowland Scots and northern Cymberi, on the retreat of the Roman power. Hence Jacket, the fmall or fides covering; Jail, or Gail, a fhut place; Jakes, the lower water fhut; January, on the fhut; Japan, on the covering part; Jar and Jargon, on the border; Jaw, the inclofing fpring; Ice, low or fhut; and Ken, Keen, and many more names under the letter K, fignifying the inclofing water edge, the I fignifying here nothing more than the article *the*.

IDA, an high hill near Troy, the high part or water part. Idell, Not. the divifion or length fpring water place. Idelton, the Idel town. Idumanum, J.ff. the divifion fpring part, haven, manfion or black water part. Jedburgh, or Gedeburgh of the Gadeni, Scot. the inclofing water fide burrow. Jena, Gena, or Ken, Scot. the inclofing water or its mouth. Jerby or Arbeia, Cumb. the water part or habitation. Here the Bercarii Tigrienfes, the Britifh or water part border keepers and the Dalmatian cohort, the divifion water fide cohort, levied amongft the Meatæ, were garrifoned, as well as at Chefter on the wall in N-hum. Thefe were afterwards, agreeable to the Roman policy of fetting the northern people to guard the foutherns, and the foutherns over the northern marches, and the fubfequent practice of the Normans, removed to the country of the Ifceni, and the levies fuppofed to have been made in Dalmatia, were made here, and our dales derive their defcent from no other country, Dalmatia being an appellative compounded of Dale and the Meatæ of Britain. Jermin, the water edge. Jern, Scot. Ire. the inclofing water, or on the water. Jet d'eau, a fpring of water. Ilford, Hants, the way or water part way. Ila, Scot. the water place or ifland. I-bre, Chef. the ifland flowing water part. Ilchefter, Som. the water place city. Ilfarcomb, Devon, the flowing water place part vale. Ilkeley or Olicana, Yorkf. the inclofing water place confines. Ilion or Troy, on the weftern water place. Illyricum or Dalmatia, the upper water place, vales or commots. Luutus, the fide flowing water place, or an hermitage. Hence Idea, the water part feen or an image; Identity, the feen fides or properties; Idiom, the furrounding fides or properties; Ides, the lower divifions or days; Idiot, from or without ideas; Idle, the fide water place; Idol, on the water place fide or image; Jealous, fhutting the fpring water place; Jelly, the fhut water place; Jeopardy, the fhut or border part; Jerk, the inclofing water part confines ; Jeft, the lowering or leffening fide, part or thing ; Jet, the water part ; Jewel, the fpring water place ; If, the fpring or flowing part or life ; Igneous, the upper confines fpring ; Jilt, fhutting the fide water place ; Ingle, the inclofing water place; Ifle, on the water place or an alley; Iliac, the ifle or alley of the bowels ; Ill, without flowing ; Illegal, without law ; illude, without light or colour; illumine, the furrounding light in; Illuftrate, the light at the lower part.

IMBRUS, an ifland of Thrace, dedicated to Mercury, the furrounded water part. Imanuentius, who was flain by Caffibeline, chieftain of the Wenta Ifcenorum part or Idumanum. Ing or Inch, Menas ing, Marget-ing, Frier-ing, and G-ing-grave, Eff. Scot. the inclofing water or ifland confines, parts, mere gate, water part, and warden. Ingelborn or Malmfbury, Wilts, the inclofing or furrounding water place ifland or English neighbourhood. Ingerfton or Ingeftre, Eff. a town on the inclofing water fide. Ingleburrow-lefield-lethorp-leton-lewood-lifton, Lanc. Berks, Norf. Bucks, Cumb. Scot. the inclofing water place, ifland or Englifh burrow, dwelling place, water gate part, town, wood, and fide town. Innerlothy-neffe-nethurie, Scot. on the inclofing water place fide, promontory, on the fide and fpring water. Infubres, Italy, the inclofing fpring water parts. Induciomarus, a chieftain of the Treviri, the mere fpring water part confines chieftain, or the Rhine grave or mere grave or warden. Hence Image, the furrounding water edge, rim, or fhadow ; Imbibe, the furrounding water part fpring part ; Imbody, furrounding the fpring part ; Imburfe, furrounding the fide fpring water part ; Imitate, at the rim fides ; Immenfe, the lower furrounding part ; Immortal, on the furrounding fea or orb fide ; Immure, the furrounding fpring water or wall ; Imp, the rim or border part; Impair, the rim or border part ; Impede, the rim part fide, which impedes ; Imperial, the high furrounding water part ; Import, the furrounding water part or port ; Impudent, at the furrounding fpring part fide ; Impure, on the furrounding fpring water part ; Impute, to the furrounding fpring part; In or an Inn, inclofed or within fome place ; Inable, the inclofed fpring place ; Inanimate, the mind fhut; Inane, without exiftence ; Inbred, the inclofing water part fide, or within the border; Incarnate, the inclofing border on the fide ; Inceft, inclofing the fame fide or border ; Inch, the inclofing edge ; Incifion, the inclofing water fides, which are cut; Incite, inclofing the fide ; Include, in the inclofing water fide ; Increafe, in the lower border ; Induct, at the inclofing fpring water part confines;

fines ; Inert, without a fpring part ; Inexpert, without being at the water part confines fide ; infant, the fpringing in ; Infufe, inclofing the fide fpring part ; Ingot, inclofing the infide ; Ingrefs, within the border fide ; Inhabit, abiding within ; Inherit, within the fide fpring ; Initial, on the infide ; Inland, within land ; Inlet, let in ; Inner, on the inclofing water ; Inroad, a road within ; Infide, within fide ; Inter, in the fide water, which is between the two fides ; Into, to the fide or within the border ; Intrench, inclofing the water fide confines ; Intrude, to the inclofing fpring water fide ; Invade, at the inclofing fpring ; Invalid, the inclofing fpring water place fide ; Invite, to the dwelling ; Inure, on the fpring ; Inward, within the fpring water fide ; Inwrap, inclofing the fpring water part.

IONIA or Iona, of Afia Minor, Peloponnefus or Morea, and an ifland of Scot. on the weft or lower border ; io fignifying the weft, io-n or ionia, the fun in the weft or on the weftern border. Jordan, on the border water part. Jovanum, in Noricum, the weftern fpring part. Ipfwich, Suff. the fide fpring part confines, ftreet, or fortification. Ipres, the lower water part. Irchenfield, Heref. the upper inclofing water dwelling place. Irk, Lanc. the upper confines. Iret, Irton, and Irtindale, Cumb. the fide water and the fide water town and vale. Irwell, Lanc. the fpring water place. Irwin, Scot. the inclofing fpring water. Ireland, Ierna, Hibernia, Hiberia, Iris, Iernis, and Bernia, the water land, the inclofing, fide or lower water part or border, or the ifland in the high water part. Ireland, though it may participate of as great variety of individuals and dialects as any European country, feems in a national view only diftinguifhable into two forts of people and languages, namely, the old Irifh or Britifh, and the Gaulifh, Welfh or Englifh, whom the Irifh call Fingallians, Englifh, Welfh, and foreigners ; or in other words, men inhabiting the fea coafts and the dwellers in vales or midland countries, as the following explication of the names of their clans or fepts, to which the O, Mac and Clan, to denote their origin and poffeffions have been prefixed, fufficiently evince, viz. O and Mac Byrn, Bryn, Bren, or Brien, on the water part or Britifh circle, chieftain or judge ; and Mac Baron and Clan Boy, on the furrounding fpring water part baron ; O and Mac Conel, Coner, Cartan, Carol, Canna, Coghlan, and Conghir or Con-more, the fea or furrounding water fide, bank or circle, offspring or chieftain ; and Canna, Coner, Dun or Roo, Cowley or Hawley and Cloyd or Floyd, the inclofing fpring water place circle, offspring or chief. O or Mac Dempfey, Dermot, Dermond and Dogherty, the furrounding water part or fea circle

circle, offspring or duke, and O, Mac or Clan Donel, Donel Tolan or Irish Donel, Dono, Dounel, Dole, Donagh and Decan, on the division spring water place confines, circle, offspring or lord. O or Mac Faly, Ferel and Feogh, the sea circle or side part man or bailey, and O Ferivil or Fervile, the spring water place or vale dwellings or dwellers. O, Mac or Clan Ge, Ghar, Genis Griffin, Granels, Gre, Glanchies, Gilaspie and Gilpatric, the inclosing or sea water confines, country or race, and McGuilly and Guire, the spring water place confines vale or offspring. O Hagan, Harragh and Harris, on the high water side circle, offspring, or warden, and O Hanlon and McHyle, on the spring water place side or bank circle, or offspring. O Jores, the sea shore circle or offspring. O Kelley, the inclosing spring water place, cell or grove circle or offspring. McLoghleim, Loghlen, and Loghton, the lake or inclosing water place brim, bank or town circle or offspring. O and Mac Molaghlin, More, Moile, Morris, Morog, Morran and Morthe, the surrounding water or sea side bank, or confines circle or offspring, and O and Mac Murray, Murgh and Mahon, the surrounding spring water confines circle, offspring or great race. Mac Nemar, on the sea circle or offspring; and O Neale, Nel or Neile and Nolan, the spring inclosing water place or bank, circle or offspring. McOfpae, the sea confines part circle or offspring, and McOwim, the spring border circle or offspring. McPherson, on the water part side circle or offspring, and McPhelim and O Phaly, the spring water place rim or vale circle or offspring. O Quin or Guin, the inclosing spring water place circle or offspring. O Felan, Ricard, Realy or Reily and Rork, the inclosing spring water place side circle or offspring. McShaglin, Surley, and Swiney, the side or inclosing spring water circle or offspring. And O-wen and McWilliam, the inclosing and surrounding spring water place circle and offspring. Other nations have been accustomed to the like distinctions by ap or ab, ac, ben, fitz, van and fon, and even the names of the patriarchs are compounds of Ab-haram, from or the son of Haram, Ha-ac, the lower country race or offspring, and J-ac-ob, the son or or from the offspring or race, and the like particles are incorporated in most of the English names. The Irish writers, it is true, derive the original of their nation from six different sources, namely, Noah's grand daughter, Cesarea by name: Bartholin, a Scythian; Nemetha, the son of Aguonan, another Scythian; Delà, with the Greeks; Gaothel and his wife Scota; and Hiberus or Hermion and Ever or Erimon the sons of Milesius out of Spain. But, however probable from the Cantabrian vocables to be met with in the Irish

language,

language, which do not occur in the Welsh, it may be that some Spanish, Scottish, Saxon, or Scythian colonies passed from thence by sea into Ireland without any great intercourse with the Britons of this island, it is still evident that those very colonies must have been of the same origin as the old sea Britons, Greeks, and Germans of Spain, Amorice, and Britain, though not wholly of the Cymbri, Gaulish, or English race, the inhabitants of vales, inland countries and cities, who were the original possessors and planters of the most considerable inland parts of both islands, and whose dialects, as well as people, are incorporated with the Irish, and into one or the other of these two, those six distinctions of the Irish writers seem to be reducible; for Cæsarea means nothing more than the lower water country, the island or Ireland : so that that colony might be from the Isceni or Ireni of Britain ; Hiberns and Hermion, the lower or western sea parts, or the Mounster men from Cornwal and Wales ; Bartholan and Nemetha, the water parts or river bank coasters, as having coasted with the Britons, Calledonians, and Gauls, of our island from Armorica ; and Dela with the Greeks or Gaothel and his wife Scota, a mixture of the former with the Cymberi-Gauls or English, inhabitants of the vales, to whom the Welsh have given the appellation of Guidels, Guidel glas, Guidel Fictiad and Verdons, all signifying dwellers in vales or hills and dales ; and some Danes and Normans have settled among them. The names of the Irish Pentarchy express nothing more than their situation, and those of the people are nearly the same as the English, without the Irish prefixtures, and they may perhaps serve to illustrate a draft of ancient Britain in its infancy. Hence Jog, Join, and Joke, the inclosing circle ; Jointure, the inclosing circle land ; Journey, on the inclosing water circle ; Joy, the circle or the sun ; Ire, the fire ; Iron, the fire one ; Irony, from the truth ; Irradiate, the rays at ; Irrational, from rational ; Irruption, the springing up on the side.

Isanium, a promontory of Ire. on the side or lower part. Isanavaria, N-Wales, the lowest in the north part. Isca, Devon, the lower inclosing water. Isca, Dammoniorum, the lower inclosing water of the Dammonii or western sea side. Isis, the lower Isceni or island side. Isis, the side or lower water place or an island. Isham, Camb. the water place side or island vale. Isleworth, Midd. the side spring water island or by the island. Isip, Oxon, the side or lower spring water place part. Ister or Danube, the lower side or division water part. Isurium Brigantium or Aldburrow, Yorks. the side lower spring water of the Brigantes. Itchen, Hants, the inclosing clofing water or Kentifh fide. Itene or Iden, Bucks, the fide or boundary foreft or vale. Ithanchefter or Othona, Eff. a city at thefide water. Itium or Portus Itius, Iftius, Iclius or Calais in Belgic Gaul, or Flanders, the lower, fide or ifland water place confines or port. Ivel, Dorfet, the fpring water place or the well. Ivelchefter, Som. the fpring water place city. Iverton, N-hum. the fpring water town. Ives, Hunt. the fpring fide. Julia Strata, the fpring water place or Severn ftreet. Juno, on the furrounding fpring. Jupiter, on the furrounding fpring part. Jura, an ifland, the fpring fide water country. Jula, the water or fpring water fide. Judea, the water part or fpring water part fide. Hence Iffue, the flowing fpring from below or the fide; Iftmus, below the furrounding fide; It, the fide, part, poffeffion or pro-perty; Itch, the fide ake; Item, the furrounding fide; Itinera-ry, on the inclofing fide; Jubilee, the fpring water place; Judge, the fpring edge; Jug, the fpring confines or fhut; Juggle, a fhut on the fpring of light; Juice, the fpring fide; July, the fpring of light; Jumble, the furrounding fpring place; Jump, the fpringing about part; Junction, the fide fpring water inclofing; June, the fpring in or the high fpring; Jury, the fpring of truth; Juft, the fpring fide part or property; Jut, the fpring fide; Juvenile, on the fpring part.

KAE, a part inclofed by water or a field. Kaer, an inclofed place, dwelling or city by walls ditches or political union. Kantre, Wales, an hundred towns or tythings of ten freeholds each, the firft Celtic divifion of countries and models of go-vernment in commots or cantons. Karn, Wales, on the in-clofing water, a bank or an hill. Karneth, Wales, an incle-fure of fortification on a hill. Kafteil, Cafter, or Caftle, Wales and England, an inclofure or fortification on the fide water place. Keeps, Yorkf. the inclofing water part fide. Kelian, N-Wales, the inclofing water place bank. Kelley, Wales, Ire. on the inclofing water place, a cell, a covered place or grove. Kelhop, Dur. the inclofing water place high foring part. Kelfay and ——— ——. Scot. the inclofing water place or fea fide or cell. Kendles, Wales, the inclofing water place fur-rounding fide or champain country. Kemburrow and Kempf-ford, Gloc. the furrounding water part fide burrow and ford or way. Kemfey, Worc. the furrounding water fide. Kemf-ley, Kent, the furrounding water fide place. Ken or Can, W-mor. Devon, Galloway, the inclofing or foremoft water confines. Kenchefter or Ariconium, Heref. the inclofing water confines or foremoft city. Kendale, Candalia or Con-cania, W-mor. the inclofing water fide vale. Kenelworth,

War. the inclofing water place fide fpring or dwelling part Kenarth, Wales, on the inclofing water fide. Kenet or Cunetio, Wilts, the fide fpring water. Kenford, Suff. the inclofing water or Ikeni ford or way. Keningha l, No f. the inclofing water or Ifceni confines hall. Kenington, Surry, the inclofing water confines town. Kent, the inclofing water or foremoft coaft, foreland, promontory or canton of a hundred tythings of ten frecholds each ; the Ifceni being the fecond and Trinobantes the third cantonment. Kenuith, Devon, the inclofing water foremoft part or premontory. Kerr, Scot. on the border or inclofing water. Kerry Wherry, Ire. the inclofing fpring water borders. Kefteven, Linc. the lower inclofing fide part, coaft or haven. Keifton Heath, Kent, on the lower inclofing water heath. Kefwick, Cumb. the inclofing water fide wick, ftreet or fortification. Keterick or Catarractonium, Yorkf. the inclofing fpring water fide confines, key or fhut. Keith, Scot. the inclofing water fide or the key. Ketering, N-ham. the inclofing water confines, gate, key or city. Ketleby, Linc. the inclofing water fide place part or habitation. Kew, Surry, the inclofing fpring water confines or key. Hence Kalends and Kalendar, on the inclofing water place bank fide or the fore divifion of time as the beginning of a month ; Keel, the water covered place ; Keen and Ken, the water edge ; Keep, the firft part or inclofing water part ; Kennel, a fhut in place or water place ; Kernel, in a covering place ; Kerfey, the fea border, which is courfe ; Kettle, a water place fhut ; Key, the inclofing water or border ; Keyage, the key edge or gage.

KIDDERMINSTER, Worc. the inclofing water part fide monaftery. Kil-cumbright-dale-dare-drumy-fenerag-flos-kenkeny-keran ley-aley or lalo-air-lin or lan-mainam-more or murry-nfey-pee-ton, Scot. Yorkf. Ire. Wales. N-ham. Heref. Dur. the fide or inclofing water place vale confines part, vale, divifion water, or water part, furrounding fpring part, place or bank, on the water part edge, flood fide, confines, on the border, place, alley, country, bank, furrounding part, fea or r ver, on the fea, confines part or town, and Chilton and Hilton by dropping the C. Kime, Linc. Der. N-Wales, the furrounding water or fea confines bank or key. Kimbolton, Hunt. the furrounding water bolt town. Kinbary, Gloc. the inclofing water fide or border bar. Kincairn, Scot. on the inclofing or foremeft border. Kin or King card n-derten eterfley-fare-horn-loffe-rofs-ton or flen-tor-thury-weil, Scot Chef. Yorkf. Heref. Staff. Surry, War. Wales, Berkf. the inclofing water border fide, part town, fide, fide place, l ar, high fpring,

7   lower

lower or fide place, border fide or morafs, town, tower border or country, dwelling part or burrow and fpring or dwelling place. Kirby Thor, W-mor. the inclofing water part or
border gate or Thor's border pa t. Kirkby-ham-intiloch-ley-
by Thor-cowbright-ton-l.fton, Eff. Leic. War Yorkf. Lanc.
Linc. W-mor. the inclofing water or border confines part or
habitation, vale or home, on the lake fide, place, gate part,
fpring water part confines, town and lower place town. Kir-
fop, Cumb. the border lower part. Kirtling or Cathidge,
Camb. the inclofing water place border. Hence Kid, kidney
the twins; Kilderkin, on the water part fide veffel; Kill,
fhutting the flood of life; Kun, on the fire fhut; Kimbo, the
fur ounding water part; Kin, in being, generating or inhabiting together; Kind, the kin fide. Kindle, generating fire;
King, from Ki-ng, chief of the inclofing water or Englifh
confines, and not from Cyninge, as per Camden; Kingdom,
the limits, borders or dominion of a king; Kirk, the inclofing
water confines, meeting or church; Kifs, the fides or the
fexes and fex together; Kit, Kitchen, Kitten, from being together, at the fide or the common inhabited part.

KNAPDALE, Scot. the inclofing water part vale. Knap-
toft, an inclofing water part or covering border part fide.
Knarefburrow, Yorkf. on or above the inclofing water part
fide burrow. Knath, Linc. the inclofing water fide. Kneb-
worth, Herts, the inclofing water part fide fpring or dwelling
part. Kington, Kn veton or Kineton, Der. War. S-Wales,
the inclofing water fide or gate town. Knock Fergus, Ire.
on the inclofing water confines bank or rock. Knocking,
Salop, the inclofing water confines or rock. Knoll, Kent, on
the inclofing water high place or an hill. Kottingley, Yorkf.
a place on the inclofing water fide. Knotsford, Chef. the
border meeting ways. Hence Knap, a fwelling part; Knap-
fack, a fwelling part fack; Knave, a covering part; Knead,
fhut or inclofed; Knee, a joint or fhut; Kneel, on the knee;
Knel, the knee, fhut or kirk calling; Knife, the edge thing;
Knight, at the inclofing water confines and an horfeman in
a fecondary fenfe; Knit, inclofing fides; Knob, a fhut part;
Knock, Knot, Knuckle, a fwelling, fhutting or inclofing.

KOLWIN or Colwen, Wales, on the inclofing fpring water
place confines or a colony of the Veneti, Wents, Venedotians,
or dwellers on the fpring water fides or in vales. Konftolen,
Confeolen or Roll rick, Oxon, the inclofing fpring water place
border, bank, pillar or kingdom, roll of the feveral diftricts
therein, which alfo ferved for mere ftones and the temples of
the god Terminus, and of which there were many; as R o-

rick near Oxford, Stonehenge on Salisbury Plain, and West Kenet, Wilts, those at Wells, Burroughbridge, and Sterlingshire, and in many other parts of Wales, Scotland, Ireland and France, and those erected by Jacob and Laban and other patriarchs, and the Romans. Kraig Ferwin, N-Wales, the high borders on the inclosing spring water. Kromlech, N-Wales, the surrounding water or border stone. Krygfrin, N-Wales, the hill borders or stone heaps. Krvg y dyrn, S-Wales, the inclosing spring water part or kingdom border or mere stones. Kuldees, Scot. the inclosing water place side, cells and cellmen. Kum or Combe, Wales and West of England, the surrounding spring water parts or vales. Kummod or Commot, Wales, the vale residence or dwelling place. Kydweli, S-Wales, the inclosing spring water place side or a joint dwelling place. Kyn, Wales, the inclosing water place confines or chieftain. Kynet, Wilts, the inclosing water place confines side. Kynta, Kynt or Kent, the inclosing water or foremost coast. Kynwy or Conwy, N-Wales, the inclosing spring water confines. Kyle or Kylan, Wales and Scot. the inclosing or foremost water place or bank.

LABERUS Durus, Kent, I e. the spring water place part gate. Lac, Lanc. Wales, a lake or an inclosed water place. Lacock, Wilts, the inclosing spring water place confines, shut or castle. Lactodorum, Lactovodum, or Stony Stratford, Bucks, the inclosing spring water place gate, station or street. Ladan, Berks, on the side spring water place. Laddeni, Scot. on the side spring water place or vale. Lade or Ladian, Wilts, on the water place side or country. Langham, Surry, the inclosing spring water place or home. Laire Merney, Ess. on the mere spring water place. Laird, Scot. on or over the side spring water place or border. Lais, an harlot of Corinth, the side spring water place. Lalam, Midd. the surrounding spring water place, place or palace. Lambey, a British island, the surrounding water place part. Lambeth or Lanbeth, Surry, the surrounding or inclosing spring water place part side, bank, lane, street, or the Sulloniacæ or Watling street. Lamburn, Berks, on the surrounding spring water place part or abode. Lamer moor, Scot. on the surrounding spring water place moor. Lamerton or Lamberton, Devon, Scot. a town thereon. Lan, Lang, Lon, Clan or Glan, the inclosing water place, bank, lane, street, vale, yard, church yard, church or village. Lanbadern, S-Wales, a bank, church, or village on the water part. Lan-beris-boidy bren-caster-ham-daf-derden-dewi brevi-elwy-edern-ercoft-erick-ganum or lyn-ger-gerfort-gerston-

gerſton-gill-hall-ley-ho-port-ſide-ton or thony-heath-hidrock-idan-ruſt-villing-wenog-witham-yken, Wales, Lanc. Dur. Suff. Salop, Cumb. Scot. Nott. Eſſ. Wilts, W-mor. Herts, N-hum. Som. Gloc. Heref. Camb. Corn. the bank, yard, church, or village of the ſpring water part ſide, border ſpring water ſide, the ſpring water part or hill, caſtle or city, vale or home, diviſion ſpring part, ſpring water part ſide or vale, diviſion ſpring water part, ſpring water place, on the water part, water coaſt, confines water, incloſing water, border, border fort, border ſide town, incloſing water place, high water place, water place, high ſpring, of the water or port, ſide, town, heath, along the rock, the diviſion water part, the river lower ſide, the incloſing ſpring water place, ſpring on the ſea, vale ſpring ſide, and incloſed water or lake. Lara, the mother of the two Lares the field gods by Mercury, on the ſpring water place. Lares, the ſide ſpring water place. Largi, N-hum. the ſpring water place confines or country. Latham, Lanc. the ſpring water place ſide vale. Latimer, Bucks, the ſurrounding water or mere ſide place. Latium, or Italy, the border place coaſt. Latobrigi, Switzerland, the Italian ſpring water part confines or borders. Lavan, N-Wales, the ſpring water place part. Lavatne or Bowes, Yorkf. on the ſpring water place ſide. Lauden or Lothien, Scot. on the incloſing ſpring water place ſide. Lauder and Lauderdale, Scot. the incloſing ſpring water part vale. Laughton, Suff. Linc. the incloſing ſpring water confines town. Lavington, Wilts, the incloſing ſpring water place town. Lanſton or Lanceſton, Corn. a town on the incloſing water place or bank. Laxton, N-ham. the ſpring water place confines town. Layton and Laywell, Yorkf. Eſſ. the ſpring water place town or dwelling. Hence Labor, on the border ſpring water; Lace, the ſide ſpring water place; Lacerate, the ſide ſpring water place diviſion; Lack, the ſpring water place ſhut; Laconic, a cloſe or ſhut ſpring of call or ſpeech; Lad, at the ſpring, or the ſpring ſide; Ladder, Lade, Ladle, and Lady, as being at the the ſide water or dwelling place; Lag, the ſpring ſhut; Laird, or Lord, on the incloſing water place ſide; Laity, the country folks; Lake, a ſhut water place; Lamb, the ſurrounding ſpring part; Lame, a ſhut upon; Lament, the ſurrounding ſpring water part ſide; Lammas, the ſurrounding ſpring lowering, or at maſs on the firſt of Auguſt; Lamp, ſurrounding the light part; Lance, on the water place confines; Lancet, on the ſpring water place ſide; Land, on the water part or bank ſide or diviſion; Lane, on the ſpring water place bank ſide; Language or Linguage, the flowing ſpring of the voice tongue; Languith, the flowing ſpring low; Lank, the water

ter bank; Lantern, on the water side bank; Lap, the water place, art; Lapfe, below the water place part; Larboard, the lower board; Larceny, on the incloſing water place; Large, on the water place confines; Larum, on the ſurrounding water place; Lath, Laſs, and Laſt, the letter or lower ſpring water place; Latch, the ſpring water place ſide; Late, Latent, and Lateral, the ſpring water place ſide; Lath and Lathe, the diviſion ſpring water place or ſides of three or four hundreds, or wapontaches; Lathreve or Trithingreve, the lath warden; Lather, Latitude, and Lattice, on the ſpring water place ſide; Laud, the high ſpring part; Laugh, ſpringing up; Laviſh, leſſening the ſpring water place; Launch, on the ſpring water place confines; Laundreſs, on the ſpring water place ſide; Laura, on the ſpring water place or a cell; Law, on the ſpring water place; Lawn, on the ſpring water place; Lax, the ſpring water place confines; Lay, the ſpring place or ground; Lazzi, Frilingi, and Lithingi, the Saxon diviſion of the people into the common or lower people or labourers, the freemen and noblemen, or the lower ſervants or vaſſals, the free or water bank holders, and the vale holders or poſſeſſors.

LEA, Herts, the ſpring water place. Leach and Leake, Bucks, the incloſing ſpring water place. Leam, N-ham, the ſurrounding ſpring water place. Lamington, War the ſurrounding ſpring water place confines town. Leander and Hero, ſuppoſed to be two lovers of Greece, the ſpring water place on the water part, and the ſea water or a river diſcharging itſelf into the ſea, or their ſimilar heroglyphick parts. Leeal, Ire. the incloſing or incloſed ſpring water place. Lechlade, Gloc. the incloſing ſpring water place ſide or country. Lechham Wilts, the incloſing ſpring water place vale, home or dwelling place. Lechamſted, Bucks, the incloſing ſpring water place vale, ſeat, or ſtation. Leckingfield, Yorkſ. the incloſing ſpring water place confines dwelling place or field. Ledbury, Heref. the ſide ſpring water or dwelling part. Lediard, Som. Wilts, the part or ſide on the ſide ſpring water place. Ledſton, Yorkſ. the ſide ſpring water place ſide town. Le, Lee, Ley, Lees, Leir, Leeds, and Lethe, Leic. Cheſ. Wales, Liſſ Kent, Yorkſ. Greece, the ſpring water place ſide, houſe, leet, or palace. Leefwood, N-Wales, the ſpring water place ſide wood. Leerpole or Leverpole, Lanc. the ſpring ſurrounding water or poole place. Legeolium, Legolium, Kirkby, Pontefract, or Caſtleforth, Yorkſ. the incloſing ſpring water place or border ſide ſhut, fortification, broken rivers or wapontaches, or caſtle way. Leon or Caer Leon, Wales and Cheſter, a city on the incloſing ſpring water place or a colony.

Legia,

Legra, a river at Leicester, the inclosing or border spring water place, or the vale border water. Leibourn, Kent, the spring water place neighbourhood. Leicester, Legicester, Leogora, or Loegora, the spring water place confines, borders or vales, or the city of the vale inhabitants, who divided their countries by rivers into wapontaches or hundreds, under the direction and government of the earl and baron, after the manner of the Gauls, Cumberi, and English, but the Britons were mostly sea coasters. Leider, Leiger, Leigh, and Leik, Yorkf. Staff. Worc. Scot. France, the inclosing spring water place. Leinster or Langenia, Ire. the inclosing spring water place side division or country. Leith, Scot. the side water place. Leiton or Durolitum, Eff. the side water place gate or town. Lelanonius, Scot. on the spring water place bank. Lemanis Portus, Limen, or Lime, the water place side, port or haven. Lemington, War. the surrounding water place confines town. Lemster or Leominster, Heref. the surrounding spring water place edge, station, minster or monastery. Len or Lyn, Norf. on the inclosing or inclosed spring water place, bank or pool. Len-ham-ox-ton, Kent, Scot. Nott. the inclosing water place vale town or upper bank. Leod and Natan Leed, Hants, on the spring water place side districts, divisions, families or tribes. Leonards forest, Suff. a forest on the inclosing spring water place side. Leskerd, Corn. the lower spring water place border. Lesley, Scot. the lower water place or the island. Lestoffe-tormin-withiel or uzella, Suff. Corn. the lower water place border part, edge or spring water place. Ledholow, Corn. a water place below the border. Letrim, Ire. the water place side, rim or border. Lettidur, Bedanferd or Bedford, the side spring water place, gate way, ford, station or habitation. Leucarum or Loghor, S-Wales, the border spring water place. Leucopibia or Whithern, Scot. the inclosing spring water place dwelling parts. Levels, Yorkf. on the slang water places. Leven, Cumb. N-Wales, on the spring water place or marshes. Leventhorp, Herts, the inclosing spring water place gate. Levin and Levington, Scot. the inclosing spring water place confines town. Lewes, Suff. the spring water place side. Lexington, Nott. a town on the water place confines. Leyburn, W-mor. the spring water place neighbourhood. Leyden, on the water place side. Leymouth, Eff. the spring water place mouth. Leyton, Bed. the spring water place side town. Hence Lead, the water place side; Leader, on the water place side; Lent, off the place; League, the water place confines together; Leak, the water place confines; Lean, the inclosing water place; Leap, the water; lace part; Learn, on the inclofing water place; Leafe and Leafh, the fpring water place fides or divifions; Leaft, the fpring water place lower fide; Leather, on the fpring water place fide; Leave, from the fpring water place; Lechery, on the inclofing or border water place; Lecture, on the inclofing water place fide; Led, the water place fide; Ledger, on the water place edge or confines; Lee, the place next the water oppofite to the wind; Leech, the fpring water place edge; Leer, on the lee; Lees, below the water place; Leet, the fpring water place fide, houfe, court, or palace; Leward, towards the water place; Left, the place off the fpring fide; Leg, the fpring place; Legg, the fpring water place confines; Legate and Legacy, at the water place confines; Legal, as inclofed within the borders; Legitimate, within the furrounding water place borders; Legion, on the inclofing water place; Legiflator, lord of the inclofed water place or border; Leifure, below the fpring water place; Len or Fanelhen, in German, the inclofing fpring water place bank, or the fee of a baron; Lerlen, church lands; Len Epifcopi, bifhop's land, land being a compound of len or lan, the water bank, and d a fide, part or divifion; Lend or Lent, the inclofing fpring water place fide or divifion; Length, the inclofing water place fide; Lenitive, like water; Leonine, the fpringing one; Leopard, the lion part; Lefs, the fide fpring water place; Let, the fide fpring water place, which hinders or fuffers; Letter, on the water place fide; Levant, on the fpring rifing or eaftern fide; Level, on the fpring water place; Lever, on the fpring water place; Levity, a fpringing property; Lewd, at the fpring water place.

Linnius or Liffey, Ire. the inclofing fpring water place. Lichfield, Staff. the inclofing fpring water place dwelling, or a ftony field in a fecondary fenfe. Lickey hill, Worc. the inclofing fpring water place hill. Lid, Kent, London, Suff. Camb. Scot. Cumb. Devon, the fide fpring water. Lidefdale-ford-fton-gate, the lid fide vale, ford or way, fide town or gate. Liga or Ligon, an ifland, in the inclofing water place. Liguria, Italy, the water place border. Lilbourn, N-ham. the fpring water place neighbourhood or abode. Lillefhull, Salop, on the fpring water fide high place. Lilingfton, Bucks, the inclofing fpring water place fide town. Lime, Kent, Dorfet, the furrounding fpring water place. Limerick, Ire. on the furrounding fpring water place confines. Lining, Kent, on the furrounding water place confines. Linen, S-Wales, on the furrounding water place, or the water place edge. Linn, Linc. Nott. the inclofing fpring water place bank

bank or a lake. Lindley, Leic. the inclosing spring water place side place. Lindum or Lincoln, the inclosing spring water place side town or bank. Lin-dham-isfarne-fey-gones-stock-lithguo-ternum, Yorkf. Dur. Linc. France, Cumb. Scot. Italy, the inclosing spring water place or water bank vale, lower side part, sea, confines, lower border, sea side or lake. Lifia or Lis, the lower or island water place. Lifmore, Ire. the sea side water place. Liffa, Spain, the lower water place. Lithhill, Surry, the spring water place side hill. Littleburrow, Agelocum, or Segelocum, Nott. the furrounding spring water place burrow. Littlecot, Wilts, the flowing spring at the water place confines. Littus Altum, Scot. the high water place side. Littleton, Wilts, Mid. N-ham. Der. a town on the flowing spring water place. Liver, Corn. the flowing spring water. Hence Liar, from the spring of truth; Libel, a calling part from the spring of truth; Liberty, the spring or flowing of the water part side, border or property; Lice, the water place side, or the small multitude; Licence, on the spring water place confines; Lick, the spring water place confines; Lid, the spring water place side or a covering; Lie, from the spring of light; Liege, the spring water place or border edge or chief; Lien, on the spring water place; Life, the flowing spring water part; Life, spring up the side; Light, the spring of fire property; Ligation, the inclosing spring water place on the side; Like, the spring water place confines; Limb, the furrounding spring water place part; Limit, at the furrounding spring water place; Limn, on the furrounding spring water place or the shadow; Limpid, the furrounding spring water place side; Line, the inclosing spring water place; Lineage, the inclosing spring water place edge or race; Link, the inclosing spring water place confines; Lip, the flowing part; Liquid, the flowing water part; Lifp, the lip low part; Lift, the spring water place lower or under side; Litigate, at the side spring water place confines; Litter, on the water place side; Little, the spring water place at the side water place; Live, the flowing spring; Liver, the flowing spring water.

Lorrini Hills of Phrygia, the circle or border spring water place banks or hills. Lochor, Loghor, or Leucarum, S-Wales, the border spring water place. Logia, Loch, or Lough, Scot. the inclosed water place or a lake. Loghty, Ire. the lake side. Loder, W-mor. Cumb. the circle spring water part. Lodge-lane, Der. the circle spring water place edge lane. Ledge on on the wolds, the border spring water place on the wilds. Loilham bridge. N-ham. a bridge on the furrounding water place. Lon, Lane, Long, and Lun, Lane. Wales, on the

Q   circle

circle fpring water place, bank, lane, ftreet, yard or church.
London, Londres, Longden, or Longborth, the inclofing
fpring water place fide port or town, or the bank, lane or long
town, as reaching through moft parts of Middlefex and Effex,
on the Thames fide, and an empory, key, port or ancient market or hans, under the government of a mayor and aldermen, or
the furrounding water or river border man, and his watermen.
Long-ford-obard-leat-ditch-us-fdale, Der. Germany, Wilts,
Midd. Scot. Lanc. the inclofing fpring water place confines
or bank, way or ford, the water part fide or bards, fide water
place, ditch, confines or vale. Loo or Loon, Scot. the circle
fpring water place or its bank. Leopole, Corn. the circle
fpring water place lake. Loofe and Loofeley, Kent, the circle
fpring water place fide. Lophamford, Suff. the circle fpring
water place vale, way or ford. Lorn, Scot. on the circle fpring
water place. Loffe or Lothian, Scot. the circle fpring water
fide. Loventium, Wales, the circle or border fpring water
place fide, or the Wenta or Veneti coafts. Lough-aw-burrow-
fin-lomond-nefs-rian, Scot. Leic. Ire. the inclofed fpring water place or lake fpring, burrow, on the edge or confines, on
the furrounding water place fide or mountain, promontory or
river. Louth, Linc. Ire. the circle fpring water place. Low-
ther or Lowland, on the circle fpring water place fide or bank,
fide or poffeffions. Loxa, Scot. the circle water place confines. Hence Load, at the circle water place fide; Loam,
about the fpring water place; Loan, on the circle or border
water place; Loath, at the circle water place fide; Loch, a
fhut water place or lake; Lodge, the circle water place edge
or harbour; Lofty, the circle fpring water place fide; Log, a
fhut place; Loin, the inclofing place; Loiter, the circle fpring
water place fide; Loll, on the circle water place; Lone, the
place of one, or the circle fpring water place; Long, the circle
or border fpring water place confines; Loop, a part from the
circle fpring water place; Loofe, the circle fpring water place
fide; Lop, a part of the circle or border water place; Lord,
on the circle water place fide or poffeffor; Lofe, the circle
fpring water place fide; Lot, the border fpring water place
divifion; Lotion, on the water place; Loud, the border water
place fide; Love, the circle fpring water part; Lough, a fhut
water place or a lake; Low, the fpring water place or well;
Lower, the circle fpring or flowing water place; Lowermoft;
the river at the fea fide; Loyal, on the circle fpring or border
water place.

Luceni or Kerry, Ire. the inclofing or border water place,
or the foremoft water coaft or coafters. Ludgate, London,
the

the flood or fpring water place fide gate. Ludham or Lud-wall, Norf. the fpring water place fide vale. Ludlow, Salop, the fpring water place fide place. Luffenham, Leic. a vale on the fpring water place. Luguballium or Carlifle, the in-clofing fpring water place, vale or dwelling place. Lullingfton, Kent, the inclofing fpring water place fide town. Lumley, Dur. the furrounding fpring water place. Lundenwic, Kent, a wick, ftreet, or fortification, at the fide of, or below the in-clofing fpring water place, or London ftreet in a fecondary fenfe. Lufitania or Portugal, below or at the fide of the lower water place. Lutterworth, Leic. the fide fpring water or vil-lage on the fide fpring water part. Lutetia, Lucotetia, or Pa-ris, the fide or inclofing fpring water place, part, fide or city. Lundoris, N-Wales, the inclofing fpring water place, or the pool water part fide or grove. Luton, Bedf. on the border fpring water place. Luthing, Suff. on the incloung fpring water place. Lydia, the fpring water place fides or diftricts. Lydington, Rut. a town on the inclofing fpring water place fide. Hence Lubricous, the fpring water place confines; Lu-cid, the fpring of light fide; Luck, the water place confines; Lucre, the fpring water place border; Luculent, the fpring water place bank fide; Lug, the fpring water place confines; Luggage, the fpring water place edge; Lukewarm, the furround-ing fpring water place; Lumber and Lump, the furrounding fpring water place part; Lurch, on the fpring water place confines; Lure, on the fpring water place; Lurk, on the fpring water place confines; Luft and Lufty, at the fpring water place fide; Lute and Lutulent, the fpring water place fide or bank fide; Luxation, disjointing like the fide fpring water place; Luxuriant, Luxurious, and Luxury, on the fpring water place confines; Lye, from the fpring of truth; Lymp, the furrounding fpring water part; Lyre, on the fpring water place, or Mercury's harp of four or feven ftrings, or rather his divifion of countries by rivers.

MAC, the middle, divifion or furrounding water confines, or an offspring or fon as the growth of the country, or of a mother, the country of man. Macclesfield, Chef. the middle or divifion water place fide dwelling place. Machun'eth or Maglona, N-wales, the middle or divifion water place bank. Macolicum or Malc, Ire. the middle or divifion water place fide or lake. Macon or Magon, N-hum. on the middle confines water. Magdalea, Staff. the middle divifion or great water vale. Maghertiernan, Ire. on the middle water fide Mages or Magi, S-Wales, the furrounding or divifion water confines

or chief. Magintum, Magiovinium, or Dunſtable, Bedf. on the middle water ſide or edge. Magna, N-hum. on the middle or diviſion water confines or city. Maglocun, a chieftain on middle or diviſion water place, or the Severn chieftain. Hence Mac, a ſon, or the offſpring of his mother country; Macaroon, a child cake; Mace, at the diviſion water ſide; Macerate, to ſteep in water; Machine and Make, the incloſing water wedge; Mackerel, the race of the ſurrounding water; Macrocoſm, the ſurrounding border of the world; Macula, a ſpot on the ſurrounding ſpring water place; Maggot, ſurrounded like an ingot in a mould or mouldy cheeſe; Magic, the action of the Magi, whoſe names like the Druids, ſignifies the diviſion or middle water ſide, their uſual reſidence; Magiſtrate, the Magi at the water ſide; Magnet, at the female water confines; Magpie, the great bill; Magnitude, in the ſurrounding water ſide.

MADUS, Madoc, and Medway, Kent, N-Wales, the middle diviſion or boundary water. Hence Mad, or diviſion; Madrigal, or a calling at the diviſion water part ſide, or paſtoral.

MAIDEN Caſtle, W-mor. Yorkſ. Dorſet, the middle or diviſion water ſide caſtle. Maidenhead or heed, Berks, the middle or along the middle water ſide. Maidenway, W-mor. on the middle water way or ſtreet. Maidſton or Vagniacæ, Kent, the middle ſpring ſide town. Mailor or Mailroſs, Scot. Ire. N-Wales, the middle or diviſion water place ſide. Maio, Ire. the middle or diviſion ſpring part. Maire, Ire. the middle or ſurrounding water. Maitland, Scot. the middle or ſurrounding ſpring bank ſide or land, or an extenſive country. Hence Maiden and Maidenhead, the middle ſpring or ſpring head; May, the middle ſpring; Mayor, Major, the middle ſpring man, as the mayor of London over the Thames; Mail, the ſurrounding water place; Maim, dividing the ſpring border; Main, the high ſurrounding water part; Mainprize, within the ſurrounding water part ſide; Maintain, on the ſurrounding coaſt; Maize, the large corn or chaff; Make, the ſurrounding water confines or action.

MALE or Macedicum, Ire. the middle or diviſion water place or lake. Maldon, Mealdum, or Camalodunum, Eſſ. a town on the middle or ſurrounding water place ſide. Malros, one of the weſtern iſlands of Scotland, the middle or diviſion water place. Malling, Suſſ. the incloſed water place or an iſland. Malmſbury or Ingelborn, Wilts, the ſurrounding or incloſed water place, or an iſland or an Engliſh burrow. Malpas, Cheſ. the middle water place lower part. Malvern hills, Worc.

Worc. hills on the middle or divifion fpring water place.
Malwood caftle, Hants, the middle or divifion water place,
fpring fide, or wood caftle. Hence Mal or Male, the middle
divifion or furrounding water place, either animal or territo-
rial; Malediction, commanding out of the borders; Malefac-
tor and Malice, a doer befide the middle water place; Mall,
an alley or place on or along the middle or divifion water place;
Malt, the middle water fide, the ufual place of malting.

MAMIGNOT, Kent, on the furrounding water fide. Mam-
fter hill, Der. a hill on the furrounding fpring. Hence Mam
and Mamma, as furrounding, or the country of a child; Ma-
millary, the mother's flowing water part; Mamon, an amaffing
or great furrounding water man.

MAN, an ifland, in the furrounding, middle, or divifion
water part. Mancunium or Manchefter, Lanc. the part or
manfion on the inclofing fpring fide. Mandueffedun, Manvef-
fedum, or Mancefter, War. the part on the lower fpring fide.
Mandubratius, a chieftain of the Trinobantes, the divifion
fpring water part fide, port, haven, or manfion, or the Idumanum chieftain. Manod mountain, N-Wales, the manfion
or inhabited mountain. Mansfield, Nott. the furrounding
part dwelling. Manober caftle, S-Wales, the water part
fpring or haven caftle. Manor, the circle or furrounding water part or manfion. Hence Man, in the middle fcale of exiftences, or on the middle or divifion water place or parts;
Manacles, the part fhut or inclofed, or the hands in a remoter fenfe; Manage, man acting; Mancipate, man keeping; Manciple, man keeping place; Mandarin, on the water
part, or diftrict man; Mandate, Mandrake, Manes, as fre-
quenting the river fides; Manger, the inclofing border, or the
hay or eating part; Mangle, fmall or divided parts; Many,
the furrounding parts; Manifeft, the furrounding parts feen;
Manna, the fmalleft parts; Manner, the mannor or the furrounding circle; Manfion, on the furrounding water fide or
refidence; Mantle, a covering; Manure, ufing the part; Ma-
numiffion, below the furrounding border part.

MAPLETON, Oxon, the furrounding water place town.
Hence Map, the furrounding part.

MAR or Marr, Scot. the country on the middle or divifion
water part or the mid-land or country. Marcantoniby, W-mor.
the inclofing water mere or lake fide part, habitation or canton.
Marches or Mers, Scot. Cumb. Gloc. Worc. N-thum. Shrop.
Chef. the middle water confines or fhuts, mews or limits.
Marden and Marcley, Heref. the mere fide, place, foreft, vale
or town. Margan, S-Wales, the middle, or furrounding

water

water or sea confines. Margidunum or Marketoverton, Rut. on the inclosing spring or middle confines part gate or town. Maridunum or Carmarthen, S-Wales, on the middle water or mere spring part, town or city. Markgrave, the warden or overseer of the marches. Market or Margat, Kent, Ess. Herts, Devon, the mere or sea gate or market place. Marlburrow, Wilts, the mere, middle or division water place burrow. Marnhill, Dorset, an hill on the mere or middle water. Marick, Yorks. the upper or Richmond mere or surrounding water. Marlow, Bucks, the mere spring water place. Martley, Worc. the mere side place. Marton or Marcaston, Der. on or a town on the mere side. Hence Mar and Marcid, a lake standing water and moisture; March, a month, the mere or spring division; Marches, borders, or limits ; Mare or March, the mere keeping companion of the knight or mere keeper like a dog and cat ; Mareschal or Marshal, on the mere side ; Margin, Margrave, Marinate, Marine, Mark, Market and Markfman, the mere confines or warden ; Marl, the mere place, where marl or fat clay was found; Marline, the sea line; Marquis or Marchuis, the chief of the mere side ; Marquisate, his seniory ; Marry, joining the meres or parts ; Marsh, the mere or sea side; Mart, the mere or sea side ; Martial, on the mere side ; Martingal, the fretful horse tye ; Martinmas, the spring or mere drawing in mas, month or division ; Martyr and Marvel, or the saint-like life of a marquis, dedicated for the defence of the meres or boundaries of his country.

MASERFIELD or Osweftry, Salop, the middle or division water side dwelling place or the western confines town. Masham and Mask, Yorks. the middle or division water side or vale. Hence Masculine, on or upon the middle spring water place or bank ; Mash, at the water side ; Mask, the mere shut ; Mass, the side or contents of the mere or limits ; Master, the mere lower side lord ; Mastiff, the mere side thief.

MATHRAVEL, N-Wales, the middle or division spring water place side or country. Matisco, Gaul, the middle confines water side. Hence Mat, as its rushes grow at the water side ; Match, joining the confines ; Mate, a companion or being at the same confines ; Matter, the surrounding parts or bodies ; Matrimony, the same surrounding part or body ; Matross, the least over the boundaries ; Matress, the lower ; Mature, using the mat or covering.

MAULEY or Malo lacu, Suss. the surrounding spring water place ditch or lake or a bad ditch. Maultby, Norf. by the
surrounding

furrounding fpring water places or fea fides. Maumbury, Dorfet, the furrounding fpring burrows. Mauritania, on the furrounding water or fea fide or coaft. St. Maure, on the fea fide or lower coaft. Hence Maul, on the furrounding fpring water place; Maudlin, on its bank or edge; Maw, the furrounding fpring part; Mawkifh and Mawks, the leffer furrounding fpring fhut.

MAXEY Caftle, N-hamp. the middle or divifion water fhut or caftle. Maxtock Caftle, War. the middle water border fhut or caftle. Maxwell, Scot. the middle water fhut fpring or dwelling place. Hence Maxillary, as furrounding and maxim as evident as the furrounding parts or felf-evident.

MEAGILE, Scot. the middle water place fide. Meales, Norf. the middle water place fide or banks. Meanborow and Meanvari, the middle fpring water on the fea part or neighbourhood. Mearnes, Scot. on the middle or furrounding water fide. Meatæ, Mæatæ, Meth or Boyne, N-hum. S-Wales, Ire. the middle or furrounding water fide or a champaign or lowland country. Meaux, Yorkf. the middle or furrounding water confines or fhut. Hence Mead or Meadow and Meat; Meager, the furrounding water for fea confines; Mean, low or on the middle water; Meafure, the middle water fide ufage.

MECHLIN, in the low countries, the furrounding water confines bank. Mecklenburgh, in lower Saxony, the furrounding water bank burgh. Hence Mechanic, a craftfman or maker of crafts or fmall fhips at the water fide; Mecke or gentle and make.

MEDEN or Newport, Hants, the middle or divifion water fide or a port on the fpring water part. Medefwell and Medefhamfted, N-ham. the middle or divifion water fide well and vale ftation. Media, in Afia and Thrace, the middle or divifion water fide or poffeffions. Mediolanum, Milling or Meifod, Gaul and N-Wales, the middle water confines, bank or village; the meaning of names thus varying with the nature and fituation of places. Medley, Yorkf. the middle or divifion water fide place or palace. Medway, Kent, the middle water. Medona, Ire. the middle or inclofing fpring. Medop, Scot. the middle border part. Hence Medal, from its being made of metal, which fee; Meddle or Interpofe, Mediate, Medicine, Mediocrity, Meditate, Medium, Medlar, Medley, Medullar, Meet, Meeting.

MELBORN, Cam. Der. on the fpring middle or divifion water place, dwelling or neighbourhood. Melcomb, Dorfet, Dur. the middle water place vale. Melerftorm, Scot. the middle

middle or furrounding water place or fea fide town. Melfield, N-hum. a dwelling on the middle divifion or furrounding water place. Melford, Suff. the middle water place ford or way. Melienith, Wales, on the middle or divifion fide. Melton or Middleton, Dorfet, Leic. the middle water place town. Melifont, Ire. the middle water place river fide or bridge. Melwood, Linc. the middle or divifion water place fpring fide or wood. Hence Melancholy, on the middle water place bank confines ; Meliorate, at the border water place and the Welfh melion, trefoil or on the boundary water place ; Mellow and Melt, or in the other language thefe names may come from the fun or circumambient light.

Mena, Mene-ing, Meneg, Meneu and Meney, Corn. Eff. Wales, on the middle, inclofing or furrounding water part, edge or boundary or narrow water part. Menapia, Menavia and Menevia, Ire. Wales, a part inclofed by or on the furrounding water or on the inclofing water fide or edge. Mendip, Som. the edge or border fide part or hills, as the limits of the old Britons. Mendlefham, Suff. a vale or home at the fpring water place fide on the furrounding water or fea part. Menlersford, Berks, the water place fide ford or way on the furrounding or divifion water part. Menill, Dur. on the furrounding water place or the water place edge. Menteith, Scot. the inclofing water fide or edge. Hence Member, the fpring ou the trunk, fhut or furrounding part ; Men or Man, as being therein ; Menace, the border action ; Mend, dividing the border ; Menial, on the border ; Menfes, the divifion of the boundary, and in a fecondary fenfe the divifion of the fun's round by that of the moon ; Menfuration or Meafuration, which fee ; Mental and Mention, on and from the mind.

Mer or Mere, Som. Wilts, the middle divifion or furrounding water or as fuch the fea, meres or boundaries, and alfo an inclofed water place or lake. Merbury, Chef. the middle or mere fpring part. Mercia or the counties of Glocefter, Hereford, Worcefter, Warwick, Leicefter, Rutland, N-hampton, Huntingdon, Lincoln, Bedford, Buckingham, Oxford, Stafford, Derby, Salop, Nottingham, Chefter and part of Hertford, the middle, mere or fea confines or the marchers. Merdon or Merton, Wilts, Surry, Kent, Lanc. the middle, divifion or furrounding water, mere or lake town. Mercury, the fea border ; Meregate, Kent, the fea gate or port. Mereworth, Kent, the middle water fide fpring water or dwelling place. Merifield and Merival, Wilts, War. the mere dwelling place or village. Merkin, Corn. the fea confines. Merlin, Wales, Scot. the fea bank, and in a fecondary fenfe, the poffeffors of the fea

bank and the fea race or line. Mernis, Merniconcs or Verni-
concs or Vecturiones, Scot. on the inclofing water or fea fide,
confines or poffeffions and poffeffors. Merfey, Eff. Lan. Chef.
the lower middle water or mere. Mertanbrooke, Der. the
middle water or mere fide brook. Mervinia or Merion,
N-Wales, the furrounding or furrounded water part edge, mere
or lake or the weftern meres or limits. Merwald, Heref. the
mere water place fide, bank or wall. Hence Mercer and
Merchant, as trading and marketing were carried on by fea
and at the fea fide; Mercury, the furrounding water or fea
borders of Europe, along which the Phrygians, Britons, Greeks,
Germans, and the other Titans, made their firft migrations and
poffeffed all the fea coafts from Greece to Livonia; Meretri-
cious, the middle water part confines; Mercy, the middle
water confines; Merely, the middle water place; Merry, the
middle water part; Merrythought; Meridian or mid-day;
Merit, the mere right; Merfion, in the furrounding water fide,
where infants and criminals were dipped for original fin and
actual tranfgreffions.

MESE, Gaul, the furrounding water or fea fide; Meffilia,
Gaul, the furrounding water or fide ; lace. Meffapia, Italy, the
lower inclofing or divifion part or champaign country. Mef-
burgum, Spain, the divifion fpring water part confines, dwelling or burrow. Hence Mefentery, the middle inclofing part;
Mefh, the fide divifion; Mefs, a part or divifion of food or
other thing; Meffage, acting or going to the lower border;
Meffuage, an inclofure on the middle or divifion fpring
fide.

METARIS Æfturarium or Meltraeth, Norf. Wales, the
furrounding water fide or place lower land, fands or wafhes.
Meth, Ire. the middle fpring fide or champaign country.
Mettingham, Suff. the middle or furrounding water fide, ifland
or ingles vale. Hence Met, the fide furrounding water;
Metal, from its being foft like water, and found at the fea fide;
Metamorphofe, the furrounding fea coaft below the water
part or a foffe; Metaphor, for the furrounding coaft or part;
Mete and Metre, a meafure like the middle water fide; Method, the furrounding water borders fide or a country within
the fame; Metropolis, the furrounding water part fide place or
city, or as fuch the mother city in a fecondary fenfe; Mettle,
at the middle water place; Mere, fhut up or inclofing
fpring; Mexburgh, Yorkf. the furrounding water confines
burrow.

MICHEL Grove, Suff. the furrounding water place confines
grove, or as the boundary woods were ufually very large, a

R   great

great high wood. Michelham, its vale or dwelling place. Michelney, Som in the furrounding water place or an ifland in the divifion water place. Micklewood, Gloc. the furrounding water place or great wood. Mickneint, N-Wales, the high furrounding parts above the bottom. Mickelbar, Yorkf. the middle or inclofing water place part, bar or gate. Mikefgate, Bedf. the middle or furrounding water fide gate. Hence Michaelmas, the middlemas; Mickle or great; Microcofm, the little circle world.

MIDDLEHAM, Yorkf the middle or furrounding water place vale. Middleton, the middle town. Middlefex, the middle fide water confines. Middlewich, Chef. the middle or divifion water confines fhut or inclofure. Middlehurft, Suff. the furrounding or middle water fide foreft or wood. Hence Mid, Middle and Midriff, the middle divifion part; Midwife, the fide divifion wife; Midas, a king or chief of the middle or Mediterranean fide or lower water part.

MILBERY and Milbarrow, Dorfet, Wilts, the middle or furrounding water part or barrow, and in a fecondary fenfe, as battles were ufually fought on the confines parts, the military barrow. Milford or Haverford, S-Wales, the furrounding water place way or harbour. Milhum, Cumb. on the furrounding water place or fea fide. Milthorpe, W-mor. the furrounding water place or fea gate part. Milton or Middleton, Kent, Norf. N-hum. the furrounding water place fide or middle town. Hence Milch, Mild and Milk, the flowing water; Mill, from its fituation on the middle water place; Mile, from the length of the inclofing water place; Millenium, the furrounding water place or fea bank; Milt, at the water place fide.

MIMERA and Mims, Herts, on the middle or furrounding water. Hence Mimick and Mock, acting the edge of the lips, mouth, cheeks and feature by way of grimace.

MINCHINGHAMPTON, Gloc. the confines water edge vale part town. Minhevit, Corn. on the fpring fide or dwelling place at the high furrounding water. Minialbrych, Scot. the upper borders or confines edge. Minfter, the fide water part edge. Minthen, Norf. the furrounding water bank or fea fide or edge. Hence Mimicry, making the edge of the mouth or other edge at or threatening one; Mince, infide the mouth; Mind, within the furrounding part man or me; Mine, in me or my confines; Mine, on the furrounding water part, where it was firft found; Mingle, within the fame confines water place; Minim, the water rim edge; Minion or Minx, on the lips edge or confines; Minifh, leffening the edge or border;

Minifter,

Minifter or Menifter, the border water fide or edge ; Minor, the circle edge ; Mintage, on the furrounding water coaft; Minuet, fpringing round or about ; Minute, the fpring water fide.

MIROBRIGA, Portugal, the furrounding water part confines. Hence mire, the furrounding water parts ; Mirror, from the furrounding water ; Mirth, the furrounding water fide or part.

MISSENDEN and Mifterton, Bucks, Leic. a town or vale on the middle or confines water fide. Miffenhead or Notium, Ire. a head or promontory on the lower furrounding water. Hence Mis, below or at the fide of the mere or mark; Mif-behave, to behave below the mark or amifs; Mifchance, a chance at the fide of the mark ; Mifer, one below the mark ; Misfortune, Mifhap, Mifreckon, Mifrule, mifs or below the mark ; Miffile, or the furrounding water fide place ; Miffion, on the furrounding water fide place, to which place the ancient legations moftly extended ; Mift and Mifty, below the water fide ; Miftake, mifs fhutting the bounds or limits of the coaft ; Miftletoe, the border water place mark.

MITFORD, N-hum. the furrounding water or fea fide ford, way or port. Mitton, Yorkf. the furrounding water fide town. Mitylene, in the ifland Lefbos, the furrounding water place part or bank. Hence Mite, the border fide ; Mitigate, the border fide get ; Mitre, on the border fide.

MIXON, Hants, on the furrounding water confines, as fhingles is below the ifland. Hence Mix, in the furrounding water confines. Here an ifland is called Ingles, which was the firft Saxon Name of England. But fee Ing and other places, which prove that the Saxon name England fignifies the ifland or Angles country, and that by Angles was meant the Ifceni ifland.

MODBERRY, Devon, the furrounding water or fea fide har-bour or dwelling part. Modona or Slane, Ire. on the fur-rounding water or fea fide, or on the water bank fide. Hence Mede, at the fide of or within the furrounding water or the manner of what is contained in the limits of a manor or other furrounding parts ; Model, on the mode : Moderate, at the furrounding water part ; Modeft, the furrounding water part lower fide ; Modern, on or the government of the furrounding water part ; Modify, Modifh, Modulate, and Modus.

MOEL and Mold, N-Wales, on the furrounding water place fide or an hill or barren place. Mole, Surry, an inclofed or fhut water place. Moleley, Surry, the furrounding water place fide. Hence Moiety, the furrounding water part two fides; Moift and Mole, a fhut water place; Moleft, the mole lower fide; Mollify, to infufe in the furrounding water place; Molten, on the furrounding water place fide.

Mon, Mona, Moion, Manau, Anglefey and Man iflands, in the weftern fea, in the furrounding water, the weftern furrounding water or the water part. Monaghan, Ire. on the furrounding water confines. Monk Coningfton, Lanc. the furrounding water confines fide town. Monmouth or Wentfet, a feat or ftation on the fpring fide or coaft on the furrounding water mouth. Monwy, the furrounding fpring water. Mount Edgecomb, Corn. on the furrounding water fide or mount edge vale. Montgomery, or Baldwin, N-Wales, the furrounding water fide or mount or a vale or dwelling place on the fpring fide. Mountfore, Leic. the furrounding water fide or mount lower border. Montrofs or Celurca, Scot. the furrounding fpring water confines fide, morafs or cell. Hence Monarch, a chief, on the furrounding water part; Monaftery, on the furrounding water part fide. Monday, the furrounding water part or moon divifion; Money, the mine, as being firft found on the furrounding water or fea fide; Mongrel or Mungrel, the furrounding water race; Mon, man or in the furrounding water place circle or midft; Monkey, man dog; Monody, a man ode; Monofyllable, one compound or labial found; Month, the moon divifion; Moon, high water circle or orb.

MORELAND, Staff. the furrounding water bank fide, poffeffions, more, marfhes and commons ufually left uncultivated on the boundary fides. Moravin, Scot. the furrounding water or fea fide, marfh or Foreft. Morefby or Morbium, Cumb. the furrounding water or fea fide part, habitation or ftation. Moreley, Devon, N-hum. the furrounding water or fea place or palace. Moreman, Devon, the furrounding water or fea part. Morganium, S-Wales, on the furrounding water or fea confines. Moricambe, Cumb. the furrounding water or fea crooked part or bay. Moridunum or Seaton, Devon, a town on the furrounding water or fea part. Morini, on the fea or furrounding water part or the fea coafters of Belgic Gaul, or Flanders of the fame race as the Britons of our ifland. Morilton, Scot. a town on the fea fide. Mert, Devon, the fea fide. Morpeth, N-hum. the fea part or fide part. Merva bychan, N-Wales, the little furrounding water part or marfh. Morwerith on the Irifh fea, the fide, lower or weftern furrounding water or fea. Morwick, N-hum. the fea confines fhut ftreet or fortified place. Hence Moral, on the

the furrounding water place circle or country; Morafs, the fea or furrounding water fide or lower part; Morbid, a fea livelihood or food; More, the fea or furrounding water; Moreover, over the furrounding water part; Morn and Morning, the furrounding fire or fun in; Morofe, the mouth low or fhut; Morphew, the fea fpring part or fcurf; Morfel, the mouth fhut upon; Mortal, the fea coaft place or vale; Mortar, on the furrounding water part fide or ftriking the fame.

Moseley, Lanc. the furrounding water place fide. Moften, N-Wales, Lydia, on the furrounding water fide. Hence Mofs, on the furrounding water fide; Moft, the furrounding water lower fide.

Mole, Cumb. the fide furrounding water. Motindon, Kent, the part on the fide furrounding water. Motwy, Mothwy or Mothwick, Salop, N-Wales, the furrounding or inclofing fpring fide or confines. Hence Mote, Moth and Mother, as furrounding or being inclofed; Motion, the furrounding property either of water or fire; Motley, in the fame inclofed place; Motto, covering the border.

Mucclesford, Berks, the furrounding water place fide or ford. Mountfbay, Corn. the furrounding water lower fide bay. Mounfter, Ire. the furrounding lower fide part or province. Moufehole or Port Inis, Corn. the furrounding water ifland or water place. Mounog or Maunog, Wales, the furrounding water or fpring part, marfh or turbery. Muchelney, Som in the furrounding water place. Hence Move, the furrounding fpring of fire or water; Mould and Moult, the furrounding water place fide or parts; Mounch or Munch, chewing in the mouth; Mount, the furrounding fpring fide bank or hill; Mountain, the high mount; Mourn, furrounding the urn; Mouth, the furrounding or fea at the inclofing fpring part; Moufe, the little furrounding animals; Mow, the furrounding fprings; Much, the furrounding fpring or water confines; Mucilage, the furrounding water place fide or edge; Muck, Mud, Mug and Muggy, as being at the water place fide; Mue, Muff and Muffle, as being covered.

Mul or Mula, Scot. the furrounding or divifion fpring water place. Mulucha, a river of Mauritania, the upper furrounding fpring water place. Hence Mulatto, at the Mulucha border, or a tawney moor; Mulberry, a black berry; Mulet, at the furrounding fpring water place fide; Mule, on the furrounding fpring water place or the dumb race; Multitude, along the furrounding fpring or dumb; Mumble, Mummery, Mummy, Mumper.

MUNDEN Furnival, Herts, the furrounding fpring water part fide den or vale or a vale on the fpring part. Mundefley, or Mondefley, Norf. the lower furrounding water place or fea fide. Munow, Heref. the furrounding fpring water. Hence Mundane, the part or country on the furrounding fpring; Municipal, Munificence, Muniment and Munition, as belonging to the furrounding water part or country; Mundbrech, the mound, ditch or a fair breach.

MURDOCK, Scot. the furrounding fpring confines water. Murray, Scot. on the furrounding fpring water Murtlake, Scot. the furrounding fpring water fide lake. Hence Mural, on the furrounding fpring water, a wall or fhut place; Murder, a killing by fecret ftriking; Murmur, a foft furrounding fpring; Murray, the furrounding fpring or a dark red; Murrion or a helmet, as covering the furrounding fpring water.

MUSKERRY, Kerry, Ire. the furrounding fpring water confines fide or borders. Mufgrave or Mofgrave, Cumb. the furrounding fpring water or fea fide warden. Muffelburgh, Scot. the furrounding fpring water place burrow. Hence Mufch, the furrounding fpring place or race; Mufe, the fpring or river fide or the mouth fhut; Mufic, Mufhroom, Mufk, Mufket, Muftard, Mufter and Mufty, from their various fprings; Mute, the mouth fhut; Mutilate, the fpring fide place divided or deprived; Mutiny, below the border fpring; Mutton, Mutual and Muzzle, on the mouth.

MYNWY, Wales, the edge fpring. Mynte Hill, Scot. the border edge hill. Myrwy or Thanet, Kent. the border or inclofing water fpring. Hence Myriad, the inclofing water or fea fide; Myrmidon, on the inclofing water or fea fide; Myftery, at the border water part; Mythology, a difcourfe on or the doctrine of the furrounding fpring borders or fables.

NAAS, Ire. in the lower part. Nabeus, Scot. the inclofing water part fide. Nablia, Germany, the inclofing water place or vale. Hence, Nab, inclofing part, or in a prifon; Naam, in an inclofed place, impound, or within the borders trefpaffing.

NADDER, Wilts, the inclofed, unfeen, or hidden fpring water part. Hence Nadir, the unfeen part, or point in the heavens oppofite to the zenith, or point over our head.

NAGNATA, Ire. the coaft on the inclofing water.

NAILBURNE or Vipfeys, Yorkf. the flood unfeen, or fpring up water from below. Hence Nail, inclofed in a place or unfeen.

NANT or Nampt, Wales, Chef. a narrow or unextensive vale. Nantuates, in Helvetia, the bottom or narrow vale dwellings or dwellers. Namptwich or Helath wen, Chef. the bottom falt fpring. Nappa, Yorkf. the inclofing, confined or narrow water part. Hence Name, on the furrounding part; Nap, an inclofed or fhut part, as the eye in fleep, and the joint of the neck.

NAR, a river of Italy, the inclofing water. Narbone of old Gaul, on the inclofing or upper water part, country or boundaries. Narburgh, Norf. the inclofing fpring water part confines part or burrow. Nardin or Narone, on the inclofing water fide. Narona, a river of Dalmatia, the inclofing or narrowing water. Hence Narrow, the inclofing fpring water confined by its banks, and its banks by hills, fo as to form a bottom or dingle.

NASEBY, N-ham. in the lower or fide part or habitation. Nafh point, S-Wales, in the lower point or promontory. Hence Nafal and Nafty, the nofe being prominent, and one of the moft natty or dirty parts of the animal body, which correfponds with the other parts of nature.

NATAN-Leod, the inclofing water or fea fide places, people or tribes, or the Britifh poffeffions and inhabitants along the fea coafts, from Hants to the Severn fea, and thence through South Wales. Natifo, a river of the Veneti in Cifalpine or Italian Gaul, the lower fide inclofing water. Hence Natal, within the inclofing water place or nation; Nature, the internal fpring of fides or things, as qualities are the like fprings of action in things and quantities, the parts of extenfion in motion and reft.

NAVAN, Ire. the inclofing water part. Nauphlia, in Greece, on or the inclofing water place, or a fea port. Naworth caftle, Cumb. a caftle on the fide or inclofing water. Hence Naval, Nautical, and Navy, as inclofures or fhuts on the water.

NEATH, Ned, or Nidum, S-Wales, on or along the inclofing water or fea fide or coaft. Neapolis, in Italy, Thrace, and Africa, a place or city on the inclofing water fide. Nerda near Babyon, on the water fide part. Hence Neap, the water part in; Near, on the water, as far is from the water; Neat, on the water fide.

NEBWORTH, Herts, on the fide fpring water part. Nebriffa, Spain, on the lower water part. Nebrodes Montes, in Sicily, on the water part, or water neighbourhood fides. Hence Neb, the bill of a bird, or the mouth of a river opening to the fea.

NECTAN,

NECTAN, Devon, on or below the water fide. Nectum, Sicily, in the water fide, or a neck of land. Hence Neck of an animal; Nectar, as being at firft cultivated on the promontories of Hercules, before the draining of the flat countries.

NEEDHAM, Suff. the inclofing water fide, or on the water fide vale or home. Needles, Hants, inclofed or unfeen places in the water or fide water part. Needwood, Staff. the inclofing or inclofed, or covering or covert water fide wood. Hence Need, not in, feen or exifting; Needle, in an unfeen or inclofed place, inclofed parts or things unfeen being as not exifting.

NEILF, Ire. on the water place. Neirburrow, Norf. a burrow on the inclofing water. Neirford, Norf. a way or ford on or near the water. Hence Neigh, near or on the confines water; Neighing, nigh the inclofing water place; Neighbour, the oppofite fide of the boundary river; Neither, the inclofing water, as belonging to neither fide; Neif, the inclofing water part.

NEMUS Calaterium, Yorkf. the inclofing water part fide, or Celtic grove or foreft.

NEN or Aufon, N-ham. the high inclofing fpring. Nenfield, Suff. the high inclofing fpring water dwelling place.

NEOR or Neure, Ire. the inclofing or border fpring water. Neomagus, Surry and Gaul, on the middle fpring water border or confines.

NEOT, Corn. on the circle water fide or abode. Hence Neoterick, modern, or on the fpring fide.

NEPTUNE, on the fide inclofing water or fea part. His being faid to have built the walls of Troy, alludes to the extenfion of its colonies to the utmoft limits of the ocean. Hence Nephew, a fpring within the fame inclofing part.

NERVII, Cumb. N-Wales, the inclofing water fpring dwelling place or dwellers, where the Roman cohorts were placed and had their names, as had the Nervii of Gaul, from their inhabiting the bank of the river Maes in Flanders. Hence Nerve, as inclofing the fpring of life.

NESSE, Kent, Salop, Eff. Scot. in the lower fide, or a promontory. Neffhides iflands, belonging to the Veneti of Little Britain, in the fide or lower water part or iflands. Hence Neft, Nettle, the inclofure or harbour of a bird, which were originally in holes in rocks on the fea fide, and unfeen.

NETTLEY, Hants, on the inclofing water fide place. Nethftead, on or the inclofing water fide place ftation Netherby, Cumb. on the water fide part or abode. Netherlands, on the

the water bank fide or country. Hence Net, inclofing or on the fides; Nether, the lower on inclofing or water fide; Nethermoft, on the furrounding water lower fide; Nettle, on the fide fpring.

NEVERN, S. Wales, the fpring on the inclofing water or fea. Nevile, a dwelling place or village on the inclofing fpring water, or a mere, village, or town, which were at firft along the fea fide. Nevin and Navan, N-Wales, Ire. the inclofing water or fea part or edge. Newa k, Nott. the inclofing fpring water confines, an inclofure or building on the fide water part, or a new building. Newborrow, Newburgh, or Newbury, Berks, Yorkf. N-Wales, a burrow on the inclofing fpring water part, or a new burrow or dwelling place. Newbottle, Scot. the inclofing fpring fide dwelling place. Newenden, Kent, the inclofing fpring fide wood, den, or foreft. Newnham, Devon, War. a vale or home on the inclofing fpring. Newland, Gloc. the inclofing fpring bank fide or land. Newfted, a ftation on the inclofing fpring. Newport, Bucks, a part or port on the inclofing fpring. Newton, Linc. Wales, a town, on the inclofing fpring or a new town. Newmarket, the new mere gate. New Miins, Scot. the inclofing fpring on the fea fide. Hence New, in the fpring; Never, no fpring, or the fpring inclofed, as Ever is the fpring; Neuter, the inclofing fpring water part, whofe fides are neuter; Newel, the fpringing place.

NID or Nidum, Wales, Scot. Yorkf. the fide water inclofing, inclofed, or unfeen. Niderdale, Yorkf. the Nid water vale. Nidifdale, Scot. the nid fide vale. Nidrey caftle, Scot. the inclofing water part or fide caftle. Hence Nib or Nip, a pin or part in or inclofed; Nich and Nick, an inclofed fide; Niece, in the fame fide or confines; Nigh, on the fame inclofing water confines; Night, the fire or light inclofed divifion or fide; Nightingale, the caller in the night; Nilling, no fpring of internal light, as willing is the fpring of internal light; Nimble, on the inclofing border water place; Nine, the high inclofing fide or rain; Nit, not feen; No, not on or in, as On is a contraction of io-in, the fun in, and None of no in and Noon, the fun in; Nivi Colini, Wales, the hills on the inclofing fprings and vales, or the inhabitants thereof.

NOCTAN, Linc on the inclofing water fide. Hence Noble, on the inclofing water place, and Nobility or Bonility tranfpofed; Nobody, not in the furrounding water part; Nod, dividing or marking the furrounding water part or border; Nogen, a water fhut one; Noife, the furrounding water fide; Nomades, no furrounding fides, or inclofed or fixed poffeffions; Nonne,

the

the furrounding water parts; Nome or Name, on the furrounding parts or things; Noodle, an inclofed or no light divifion; Noofe, inclofed on the fides.

NORTH or Nord, on the border water or upper fide or border. Norfolk, on the border water place confines or people. Norham or Northam, N-hum. the border or inclofing high water fide vale, or the northern vale. Normandy, France, on the border or inclofing water part fide or poffeffions, or the Norman poffeffions. Thefe people, who, from the names of the individuals, their language, cuftoms, and government, feem to be a mixtue of Danes, Germans, French, Romans, and Britons, for many centuries before their fettlement in Neuftria, had no fixed poffeffions, but like their anceftors, the Danes and Saxons, lived by their depredations in different countries. And having kept no records of their own origin and antiquities, have, like the Romans, been at much pains to darken, confound, and deftroy thofe of others, particularly the Celtic, and yet, inftead of refting their origin and antiquities on their extenfive alliance to many Celtic nations, the beft diftinction of a nation, feem to be as much prejudiced in favour of a feparate origin of their own, as any other Celtic branch, to the great encouragement of party feuds, and detriment of fociety. Normanfter, the furrounding water lower or ifland fide. Northampton, Northehamton, or Northcanton, the northern canton, or inclofing fpring water or vale confines or town. Northumberland or Chumberland, the northern Cumberl poffeffions, or dwellers in vales, or on the fpring water fides. Norwic or Norwich, the furrounding water or northern confines, ftreets, city, or fortified part. Norway, the furrounding water or border, or fea fide way. Hence Nor, the inclofing water, which belongs to neither fide.

NOSHEAD, Scot. the furrounding water lower fide or head. Noffill, Yorks. the furrounding water lower fide hill. Hence Nofe, the lower on the fide; Noftrum, an undifclofed thing.

NOTESLEY, Bucks, the inclofing water fide lower place. Notium, Ire. the furrounding water or fea fide. Nettingham, on the inclofing water fide vale. Hence Not, no fide or in: Note, a mark, or fide inclofing water or boundary; Nothing, no inclofing or exifting fide or part; Notice, warning to be at the fide or boundaries; Notify, Notorious, and Notion, remarks; Nothwithftanding, ftanding with the conclufion or inclofure.

NOVANTES of Galloway, Carrick, Kyle, and Cuningham in Scotland, on the inclofing fpring water confines or vales, or the Clyde's sides. Noviomagus or Woodcot, Surry, the inclosing or middle spring water confines. Hence Novel, on the spring, which is new; Nought, not in the inclosing spring; Novice, the lesser in the spring; Noun, on the inclosing spring parts; Now, on the spring; Nourish, on the spring side.

NUNEATON, War. a town on the inclosing spring water side. Nutley, Hants, the inclosing spring water side place. Hence Nut, the inclosed side spring of vegetation; Nudity, uninclosed sides; Null, on no spring place; Numb, no surrounding spring part; Number, on the surrounding spring parts or water parts, and for their particular definition see under Can; Numerous, the inclosed spring and sea surrounding parts; Nun, a shut or inclosed one; Nuncio, on the inclosing spring confines or border; Nuptial, inclosing or joining together the side spring water parts; Nurse and Nurture, on the spring side; Nuzzle, inclosing or shutting up.

OBOUN, Ire. the inclosing spring water part. Obris, Gaul, the side spring water part. Obulco, Spain, the circle spring water place confines. Hence Obey, the circle part; Obedient, Obelisk, the circle or border spring water place mark; Object, the circle part opposite side, or to cast from; Oblation, on the circle spring place; Obligate, the circle place gate; Oblige, the circle place edge or bounds, or bound to the circle place; Oblique, from strait; Obliterate, from literate; Oblivion, a flood on the circle part; Obscure, the circle side border; Observe, the boundary mark, or the circle side water part; Obsolete, the circle part from a side water place; Obstacle, the circle part standing water place; Obstinate, Obstruct, and Obtain, within the circle part coast; Obtrude, through the circle part side; Obtuse, the circle part lower side; Obviate, the circle part spring side, or a way along the same.

OCELLUM, Kilnsey, or Holderness, York. the sea or salt water part, promontory, or cell. Ocitania, at the sea side. Ocetis, Okitis, or the Orcades of Scotland, possessions at the side or below the surrounding water or sea side. Ochenture, Oxon on the inclosing spring water side. Ochie hole, Som. the high spring water hole. O ke, Devon, Berks, the inclosing water. Ockhampton and Okeham, Devon and Rut. the inclosing water part vale, or vale town. Ocrinum or the Lizard, Corn. a part in the sea, a promontory, or the lower water place side. Octocesa, Spain, the surrounding water lower border. Octopitarum or St. David's, S. Wales, the sea

border

border part coaſt. Hence Occaſion, the circle water ſide action; Occidental, on the water circle lower coaſt; Occult, on the water circle ſide; Occupation, on the ſurrounding water ſide part; Occur, the ſurrounding water or ſea and the current meeting. Ocean, the high ſurrounding water; Octave, the ſurrounding water part or diviſion, an eight, or an iſland; Occultiſt, the circle of light hiſtorian, liſtener, or attender.

Odcomb, Som. the circle or ſurrounding water ſide vale. Odda, Gloc. the ſurrounding water ſide or Severn part. Odiham, Hants, the ſurrounding water ſide vale. Odil or Woodhill, Bedf. the ſurrounding water or ſpring water ſide place, high place or hill. Odyſſæ Portus, Sicily, the ſurrounding water ſide lower port. Hence Odd, divided from the circle of men or things; Ode, the round dance and ſong; Odium, from the circle; Odour, from the ſurrounding circle; Oecumenical, on all the ſurrounding parts.

Offa, the ſurrounding or circle part. Offalie, Ire. the ſurrounding water place. Offington, Suff a town on the ſurrounding water confines. Offride, the ſurrounding water part ford. Offton, Suff. the ſurrounding water part town. Hence Off, the ſurrounding water part, which is commonly far or at a diſtance; Offal, caſt aſide or on the ſurrounding water part; Offence and Offend, off the fence or ſurrounding part ſide; Offer, on or bringing to the ſurrounding water part; Office, the ſurrounding water part ſide, an office being kept at its gate or port; Offing, in the ſurrounding water or ſea part; Offspring, from the incloſing water part; Oft, and Often, at the ſide of the ſurrounding water part, or the things contained therein, which are many and frequent.

Ogle, N-hum. the ſea ſurrounding water place. Oglethorp, the ſurrounding or circle water place gate part or port. Ogilvy, Scot. the ſurrounding water place ſpring ſide or dwelling place. Ogmer, S-Wales, the ſea ſurrounding water part. Ogo, Wales, a cave or a ſurrounded, ſhut or incloſed part. Ogygia, a Thracian iſland, the ſurrounding water or ſea confines or country. Hence Ogee and Ogle, a round and looking round, and Oh, the high O or the horizon, ſun, &c.

Oilwy, S-Wales, the ſurrounding water place ſpring or brook. Oiſler hil's, Herts, the ſurrounding water part ſide hills. Oiſterley, Midd. the ſurrounding water part ſide place, or dwelling place. Oiretum, Italy, the ſurrounding water ſide. Hence Oil, the flowing water; Oint, the water on the ſides, which anoints the ſides of rivers in its courſe; Oyer, on the ſurrounding

furrounding water; Oyes, the furrounding water fide; Oy-
fters, as ufually found at the furrounding water.

OLANIGE, Olantigh, Olaniage, Oleaneag, or Olenacum,
Kent, Gloc. Cumb. on the furrounding, inclofing, or divifion water bank or confines. Old-ford-ham-ftreet-wark-wyke,
the furrounding water place fide, way or ford, vale or dwelling place, ftreet or water fide part, and river confines or city,
and as places on the water fides were ancient dwelling places,
an old ford, &c. Oleron, a French ifland, on the furrounding water place. Olicana or Ilkeley, Yorkf. the furrounding
water place confines or inclofures. Oliphant, Scot. the furrounding water place bottom. Oliver, the furrounding water
place fpring. Hence Old, Oligarchy, Olio, and Olitory.

OMBRIOS, an ifland in the Atlantic, the lower furrounding
water part. Omphalium, in Greece, a place on the furrounding water. Omer, Homer, Gomer, or Chomer, the fea furrounding water part. Homebury hill, Surry, the furrounding
fpring water part hill. Hence Ombre and Omega, as the circle of all parts; Omelet, a pancake from its roundnefs; Omit,
at the fide or out of the furrounding water; Omnipotent, a
power furrounding all parts or things.

ON, No, Or, and Heliopolis, Ægypt, on the furrounding
water or border, or Io-n, the city of the fun, or furrounding
circle, the Temple of Apollo being here fituated. Onæum,
in Dalmatia, on the furrounding water part. Oneal, Ire. on
the circle or furrounding water part. Onoba, Spain, the furrounding water part. Onflow, Salop, below or on the furrounding water fide fpring water place, or a low or flow furrounding water. Hence On, from the motion of the water,
or fun; One, the circle; Onerate, at the circle water, the
ufual place of lading; Onion, a round one; Only, the furrounding water place; Onward, the furrounding fpring water
part; Ooze, the furrounding fpring water part fide.

ORBY, Chef. the border part or abode. Orcades, Orcas,
or Orkney, Scot. the inclofing water border. Ordovices, Or-
doluce, Venedotia, or North Wales, the furrounding fpring
or water parts, border, fides, confines, ftreets, or cities, or
the borders of the Veneti or Wenta, who, according to An-
gelus Capellus, Polydore Virgil, and Dr. Caius, once were
the inhabitants of Norwich, and if the pedigrees of fome of the
North Wales people are right, many of thofe families derive
their origin from that country. Oid, Scot. the furrounding
water fide. Ore, Suff. the furrounding fpring water. Ore-
ford, its ford. Oreweed, Corn. the furrounding water weed.
Ormefby, Norf. by the furrounding border fide. Ormfton,
Scot.

Scot. the furrounding border fide town. Ormond or Orwown, Ire. the border on the fea or furrounding water. Orry, Ire. on the border water. Orton, Hunt. W-mor. the border town. Orwel, Suff. the border fpring water place. Hence Oraion and Oration, en the border fide; Oral, on the mouth ; Orange, a round fhut part; Orb, the water circ e part, the fea being fuppofed to furround our globe; Orchard, an in-clofed part or yard; Ordain, on the border part; Ordeal, on the border water or avenging part; Order and Ordinance, on the border fide or poffeffions; Ore, on the furrounding water, it being formerly found along the water fide; Orient, the up-per fide border; Origin, the border water fire in; Original, on the border water fire in; Orphan, on the inclofing water or border part; Orrery, on the world; Orts, from the fides; Orthodox, the furrounding border doctrine; Orthographer, the furrounding border writer or engraver; Ora, a Saxon coin mentioned in Domefday, valued at fixteen pence, on the bor-der.

Osca, the furrounding water confines. Ofborn, on the furrounding water fide or fea fon. Ofgodby, the fea confines part or habitation. Oflan, the fea bank. Offory, Ire. the furrounding water lower border. Oftai or Oftiones, Corn the furrounding water lower, or weftern fide, weft being a cor-rupt deviation from oeft. Oftidamnii, Corn. on the weftern furrounding part. Ofwald ftreet, the furrounding water place fide ftreet. Ofweftre, Salop, the weftern furrounding water or border town. Ofwy, the furrounding fpring water fide, haven, or houfe. Hence Ofcitant, opening the mouth or fur-rounding part; Oftenfive, fhewing the furrounding parts; Oftler, on the furrounding water place fide, fhuts, or houfe; Oftrich, on the furrounding water part fide confines.

Otelands, Surry, the furrounding water fide bank or land. Othona or Ithancefter, Eff. on the furrounding water fide. Otford or Otanford, Kent, the furrounding water fide way or ford. Otley, Yorkf. the furrounding water fide place. Otmore, Oxon. the furrounding water fide moore. Ottadini or Meatæ, N-hum. below the furrounding water fide, or Tyne coaft, or the border water coaft. Ottendun, Oxon. a town below the furrounding water fide. Ottery, Devon, the fur-rounding water part. Hence Other, the furrounding water or river fides; Other gates, their gates; Otherwife, their ways; Ottoman, the furrounding water fide border man.

Over Burrow, Lanc. the furrounding fpring water part or burrow. Over-rby, Surry, en or the country en the fur-rounding fpring, or over the river. Overten or Orton, Hant.

Chef.

Chef. W-mor. the furrounding fpring water fide town. Over-thorne, Yorkf. the door or gate on the furrounding fpring. Oulney, Bucks, on the furrounding fpring water place. Oundle or Auondale, N-ham. a vale on the furrounding fpring water place or river. Ounefbary Topping, Yorkf. the upper furrounding fpring water part or bar. Ovoca, Ire. the furrounding fpring water. Oufe, Berks, Bucks, Staff. Norf. N-ham. Yorkf. the lower, leffer, or fide furrounding fpring. Oufeley, Salop, War. the furrounding fpring fide place. Oufney, Oxon. ifland in the furrounding fpring. Ouftman, Ire. the furrounding fpring lower fide, or weftern part. Out Burrow, N-hum. the furrounding fpring fide burrow. Ow, the furrounding fpring. Owen, on the furrounding fpring. Owers rocks, Hants, Dorfet, the furrounding fpring water rocks. Hence Oval, Oven, Over, and Ought, the furrounding fpring water fide or parts therein; Ounce, on its fide; Our, the fpring border; Oufl, the furrounding fpring lower fide; Out, the furrounding fpring fide; Outer, on the fur-rounding fpring fide; Outermoft, on the furrounding water lower fide; Outward, the furrounding fpring fide water part; Owe, within bounds; Outfang thief, the out part confines thief; Outlaw, out of the fpring water place border.

OXBURGH, Norf. the fu rounding water confines burrow. Oxenbridge, Linc. a bridge on the furrounding water. Ox-enhall, Dur. a hall on the furrounding water confines. Ox-ford or Oxonium, the furrounding water border, way, or ford. Oxney ifland, Kent, an ifland in the furrounding water. Hence Ox, as an inhabitant on the river confines, or the up-per kind or fide; Oxygon, three acute angles like the furround-ing water.

PABENHAM, N-ham. the vale upper part. Pas or Pace, Belgic Gaul, the lower part or country. Pachinium, Sicily, a part inclofed or fhut in the water. Hence Pace, the lower or leffer part or foot; Pack, a fhut or inclofed part or thing; Paction, a thing clofed.

PADSTOW, Corn. the part or ftation at the boundary water fide. Padus, Italy, the fide fpring water part. Hence Pad, the fide part; Padar, the afide part; Paddle, the fide part at the water place; Paddock, a part at the water fide or con-fines.

PANSUT, Chef. the part on the river fide. Paones, of Panonia, the weftern parts or ends. Hence Pæan, the wef-tern fong.

PAGANEL or Paynel, Bucks, the part on the inclofing water

ter place, or a ſtreet, town, or burrow, which were ſituated along the border rivers. Pagæ, Greece, incloſing water part. Hence Pagan, Painim, or Heathen, on the ſurrounding or border water part, the uſual dwelling places of pagans and pagods; Page, from his paſſing along the ſtreets and pageantry; Pail, the water place thing; Pain, the part in; Paint, the part on the ſide or ſurface; Pair, the water part ſides or the limits; Pagus, a ſtreet along the border river confines, as Berkſhire on the river Thames

PALATINE Counties, on the ſide of the water place, leet, or palace counties. Palladia or Toulouſe of Gaul, the border water place ſide. Palantia, in Spain, the water place bank ſide. Palus, a marſh or the water place ſides. Hence Pale, Palace, Palatinate, Paliſade, Palliate, Pall Mall, Palm, and Palſgrave, from their ſituation on or along the ſide, covering or border water places; Palate, the covering of the mouth; Pale colour, the water place colour; Palſy, ſide place; Paltry, a dry water place.

PAMBER Foreſt, Hants, the ſurrounding water part foreſt. Pamiſus, a river of Macedonia, the lower ſurrounding ſpring part. Hence Pamper, the paunch or ſurrounding part; Pamphlet, the fugitive ſurrounding or covered part.

PANCRASE, Midd. a part or bottom on the lower water ſide. Panonia, or Hungary, the Ionian or weſtern end or part. Pant, Eſſ. Wales, a part or bottom on the water ſide. Panton, Paunton, or Ad Pontem, Linc. a town on the ſpring water part, or a wapontache town. Panwen, Wales, a part on the incloſing ſpring or on the ſpring end. Hence Pan, an incloſing part or a veſſel; Panacea, as containing all medicines; Pancake, Panado, Pane, a part put in; Panel, on the incloſing water part or country; Panner, a thing incloſing; Pant, a conduit of water, or to pant; Pantile, a tile with a bottom and ſides, or a hollow tile; Pantry, the incloſing water part ſide.

PAR, the water part. Parham, Suſſ. the vale water part. Pariſii, Yorkſ. the leſſer or lower diviſion or riding. Parret, Som. the ſide water part. Hence Par, the water part or place; Paradiſe, below the water part ſide; Parallel, the water part place or ſides; Paramount, the water part mount or bank; Parapet, on the water part ſide part; Parcel, the water part confines place; Parcenery, together on the incloſing water part; Parch, the upper part; Pardon, on the water part ſide, or admitted therein from baniſhment; Part, the part upon, or the upper part; Parent, the part or head on the houſe; Pariſh, the leſſer part or diſtrict; Park, an incloſed part; Parley and Parliament,

Parliament, the talking part or houſe; Parlour, the part on the floor; Part, Partition, and Party, ſide or diviſion part of a country; Parſon, on the pariſh; Parſonage, the parſon's confines; Patron, on the ſide or border water part.

PASHAM, Bucks, the vale lower part or end. Paſheley or Paiſley, Scot. the lower or leſſer water place. Paſſelew, Salop, the lower ſpring water place. Paſton, Norf. on the lower border part or town. Hence Paſs, the leſſer part; Paſſage and Paſſion, a being ſtraitened; Paſſive, the lower part; Paſture, the lower part ſide or country.

PATERN, S-Wales, on the water part ſide. Pateſhul, N-ham. the part below the water place ſide. Patrington or Prætorium, Yorkſ. a town on the incloſing or frontier water part. Hence Pat, the part at; Patch, the covering part; Patent, on the incloſing part; Path, the ſide part, which were at firſt along the water ſide; Patient or Paſient on the lower part; Patriarch, the chief over a country or family; Patrimony, the father's part or limits; Patriot, at or on the country part; Patrol, the path roll or rounds; Patron the country part one; Pattern, archetype or ſpecimen, the incloſed or ſide water parts, which were the firſt patterns of things, and of which letters are types or repreſentations.

PAULTON, Corn. the ſpring water place town. Paveley, Wilts, the ſpring water place part. Taunton, Linc. a town on the ſpring water part. Pawlet, the ſpring water place ſide. Pawton, Corn. the ſpring part town. Paynſwic, Gloc. the part on the incloſing ſpring ſide, ſtreet, or the Wiccian ſtreets or cities. Hence Pave and Pauſe, the low part; Paw, the ſpring part; Pawn, in the ſpring part or hand; Pay, the part or ſhare.

PEAKIRK, N-ham. the incloſing water part border or meeting. Peake or Pike, Der. the high part. Peas, Kent, Suſſ. the lower part. Peaſalong, on the low water vale. Hence Peace and Peas, or a low part or thing; Peak, the high part; Pear, a watery thing; Peaſant, the lower part in the houſe or poſſeſſions; Peat, the water part ſide, or at the water or wet part.

PEBIDIOG or Pebidiog, S-Wales, and Peibles, Scot. the ebbing or flowing water part, or along the ſea ſide. Hence Pebble, on the ebbing water place part.

PECHE, the incloſing water part. Pechencourt, France, on the incloſing water part ſide, border, or court. Hence Peccant on the incloſing water part ſide; Peck and Pectoral, a ſhut part or veſſel; Peculiar, on the incloſing ſpring water

T   place;

place; Peculator, from the inclosing sp ing water place side or country.

PEDERED or Pedred, Som. the side or lower spring water part. Pedde ton, the side or lower sp ing water part town. Pedwa din, Linc. below the spring water side part. Pedum, Italy, on the spring side part. Hence Pedagogue, Pedant, and Peddlar, the lower or under part man; Pedestal, the foot or lower part or stall; Pedigree, the lower race part; Peel, the upper place; Pee, the water part; Peg, the inclosing water part.

PELF or Pell, Corn. the water place or farthest part. Pelham, Herts, the water place or farthest vale, as water places were the limits or boundaries of countries. Pella and Pela onia, Greece, the water place or farthest confines. Hence Pellhele, the inclosing place: Pellmell, the middle or division water place side, street, or strand, and Pelt, the places of hurly burly.

PEMBROKE or Penbroke, S Wales, the surrounding water part, neighbourhood or end. Pomfey, Suff. the sea surrounding part. Pen, Som. Bucks, Lanc. the inner higher or upper part, as per is the side or lower part. Penalt, N.Wales, the part on the heights or hills. Pendenis, Corn. the inner water part or vale side or lower head. Pendle, Lanc. the upper ends or hills or the interior hill parts. Penk and Penkridge or Penocrucium, Staff. the water part, end, ford, bank, street or way. Penhaw, S-Wales, the spring water part end. Penherst or Hurst, Kent, the forest lower end part. Pentre, Wales, the town's end part or suburbs. Penigent, Lanc. the upper hills on the side or the windy hills. Penrose, Wales, the part on the surrounding water or sea side or marsh. Penfans, Corn. the part on the sands. Penfavas, Corn. the part on the lower water or the water lower end. Penvael, Penwal or Walton, Sea Wales, the part on the water place or wall side or end. Penworth, Lanc. the side spring water end. Penysthorpe, Yorks. the lower end water gate. Hence Pen, the part, the inner, inclosing, or upper part or the head of a man or river; Penal, on the head; Penance, Peartence, within a pen; Pence and Penny, on the pen or head; Pendant, the head or end on the side part; Penetrate, the part entering at the side; Peninsula, an head isle or lower place land; Penthon, a shut up, fed or cooped thing; Pensive, the head low or like a slave; Pent, shut up; Penthouse or Pentice, penury; Pentagon, five angles.

PEOPLESHAM, Kent, the spring water place side or country vale. Hence People, every country place or people.

PERCLEBRIDGE,

PERCEBRIDGE, Dur. the incloſing water part bridge. Percy, N-hum. the water part confines or chieftain. Perith, Cumb. the water part ſide or ford. Perot, the water part ſide. Perſhore, Worc. the water part lower ſide or ſhore. Perſia, the water part ſides or diviſions. Perth, Scot. the water part ſide or a grove. Teryn, Corn. on or in the water part. Hence Peradventure, Perhaps, and Perchance, the high incloſing water part; Perambulation, at the ſide of the ſurrounding water part; Perch, the incloſing water or a ſhut or ſhelter part or a pole of ſixteen feet and an half in length of which forty in length and four in breadth made an acre or the border diviſion. Percuſſion, ſpringing the part; Perdition, on the water part ſide; Peregrine, by the incloſing water part borders; Perfect, the water part effected; Perforate and Perforce, to the water part border or ſide; Perform, the ſurrounding part; Perfume, the ſurrounding ſpring part; Peril, on the water part; Period, the circle or ſurrounding part or water part; Periſh, below the water part; Perjury, from the oath part; Periwig, the covering part; Permit, the ſurrounding water part; Perpetual, a part always on the ſpring; Perry, the water part; Perſevere, the divided or ſevered water part; Perſiſt, ſtanding at the water part ſide; Perſon, on the water part ſide; Perſuade, at the ſpring water part; Pert and Prig, at the ſide or mouth part; Peruſe, at the ſide or below the ſpring water part; Pert, the part below the ſide; Perdings, the farthing holders, or the dregs of the people, whoſe farthing holding was allotted him as a payment for their labour, on the incloſing water part.

PETERBURROW, N-hamp. the water part burrow. Petroc, Corn. Wales, the ſurrounding water or ſea part. Petuaria, Pariſiorum, Huldby or Beverley, Yorkſ. the ſpring water parts, diviſions or diſtricts of the Pariſh. Petworth, Suſſ. the ſide ſpring water or dwelling part. Hence Pet, Pettiſh, Petulant and Petition, the hand, foot or part at or to; Petit and Petty, the ſide or diviſion part, which is ſmall.

PEVENSEY, Suſſ. the ſpring part on the ſea or a haven. Pever, Cheſ. the ſpring part. Peyton, Suſſ. the town part. Hence Pew, the ſpring part incloſure, or an abode by the ſpring part where the firſt pews were let up.

PHELEM-ge-modona, Ire. the high ſurrounding mountain parts. Phellius, a mountain of Attica, the high part. Hence Phariſaical and Pharo, from their high appearance.

PICKFORD, Salop, the upper or high part or peak ford or way. Pickering, Yorkſ. the peake or high confines part. Pickworth, Linc. the upper ſide ſpring water part or dwelling place. Pictland, Pictavia, Phicti or Caledonia, Scotland, the

upper,

upper, high or hill inclosing spring part side, lands or possessions or highlanders, the Scots being the sea side and Meatæ the middle possessors. Hence Fight, by an inflection of the p; the Caledonians being mostly employed in military affairs; Picaroon, a robber or mountaineer; Pickle, picture and figure; Piece, a small part; Pier, the water part; Pierce, the water part side; Pike, the water part confines. On the western side of the highlands were the Caledonians, on the eastern were the Britons or Saxons called Picts, and from a mixture of those two people came the Caledonian Picts and Meatæ.

PILE of Foulney, Lanc. an high part on the water place. Pilkington, Dur. a town on the high inclosing water place. Pillerton, War. a town on the high water place. Hence Pilaster, Pile, Piller, Pillage, Pilot, as being high parts or at the high water place.

PIMPE, Kent, Wales, the border dwelling, possessed or inclosed part. Pimpleas Macedon, the upper border dwelling part lower place. Pimble, N-Wales, the surrounding or inclosed water place part or lake. Hence Pimp and Pimple, as gathering or shutting together.

PINHOE or Pinhaws, Devon, the part within the spring water or river part. Pinaria, an island in the Ægean sea, a part in the water. Pinkney, N-ham. the part in the inclosing water. Hence Pin, the part in or inclosed; Pinch, the part shutting or squeezed; Pine, an high part; Pinfold, a place inclosed or shut on the sides; Pinion, the part or pin or the trunk; Pinacle, an high inclosing place; Pinner, on the high part; Pint, the side inclosing water part or at the conduit; Pious, springing up to the high parts.

PIPWEL or Divils, N-ham. the upper spring water place part or the division spring side. Piperno, Italy, on the upper or up the water part. Hence Pip and Pipe, a springing up the part or throat; Pipkin, a boiling vessel or springing, upping or piping water shut or inclosure.

PIRANUS, Corn. on the water part. Piræus Portus, Greece, the water part gate or port. Hence Pirate, at the water part; but Pyre is the high or fire part, and Pyramid, the manner of the fire high part or flame terminating in a point.

PISA, of Greece and Italy, the lower or side part. Pisaurus, a river of Italy, the lower or side spring water part. Hence Piss, the side part; Pistol, the side part tool; Piston, on the side part; Piscatory, a fish water inclosure; Pish, a little or low part or thing.

PITCHFORD,

PITCHFORD, Salop, the upper fide water part ford or way.
Pitana, Greece, the part below, at the fide or on the coaſt.
Pithium, Greece, the upper fide part. Hence Pit, a hole;
Pitch, in a fecondary fenſe, from the bitumen of pitch; Pith,
Pittance, and Pity, the fmall part; Pivot, the pin of a round
or turning thing

PLAISY, Eſſ. the fide or lower water place, it being fituated
on the river Chelmer, which difcharges itſelf into Idumanum or
black water, and feems to be the boundary betwixt the Trinobantes and Iſeeni, who made their migration out of Kent
along the fea coaſt of Eſſex; and perhaps in a fecondary fenſe,
a pleaſant place, but pleaſantneſs is not expreſſible by the
primary meaning of names any more than taſtes, founds and
other qualities. Pianulia, Gaul, the lower or fide water place.
Placentia, Italy and Sſain, on the incloſing water place fide.
Plim, Devon, the rim or boundary water place. Plimouth and
Plimton, Devon, the Plim mouth and town. Pimlimon or
Pimlimon, N-Wales, the boundary or five flood mountain.
Plumpton, Lanc. the ſpring border part town. Hence Place,
the fide water place; Plain, on the water place; Plaſter, a
thing on a lower part; Plait, a part at the fide; Pan, on the
water place; Plant and Plantation, on the water place fide;
Plaſh, a fmall water place; Plat or Plot, a broad part or a
part at the water fide or border and platter; Play or Pleas, the
water place fide where the courts leet and baron were held;
Pleaſe, at the water place fide; Pledge and Plight, the
water place edge or fide or the giving furety of frank pledge;
Plenty, on the water place fide; Plevin, the boundary plea;
Plough, a place acting up thing; Plumage and Plumy, the
ſpringing up things; Plump and Plum, ſpring round the part;
Plunge, thut in the water place; Pluvious, ſpringing up
water part.

Po or Padus, Italy, the furrounding fpring water part.
Pole or Poole, Corn. Dorfet, Chef. Wales, furrounded water
place. Pollac, Scot. a poole, lake or furrounded water place.
Pollefworth, War. by the furrounding water place. Poltimere,
Devon, the furrounding water place fide or mere. Polton,
Corn. the furrounding water place town. Pomery, Devon,
the fea furrounding part. Pomona, an Iſland, in the fea furrounded part, Pont or Colbrooke, Bucks, the furrounding
water part fide brooke. Ponteland or Pontælii, N-hum. the
furrounding water fide bank fide or land. Pontefract, Yorkſ.
the broken rivers or border water parts. Porlock, Weced port
or Watchet, Som. the water part key, lock or port. Portland,
Dorfeſ, the furrounding water part bank fide or country. Port-
mollech,

molloch, Scot. the great lock or lock part. Pouleford, Chef. the surrounding water place ford or way. Port, the surrounding water part side. Portgreve or Portrev, the surrounding water part side or port warden. Portfey, Hants, the surrounding water part side, or port island. Pertholme, Hant. the surrounding water part side island. Portskeweth, S-Wales, the surrounding water part side or port at the lesser spring water side. Portslade, Sull. the surrounding water part side or port country. Portus Adurni or Ederington, Sull. on the water part port or watering port or the inclosing water part town. Portsmouth, Hants, the port or surrounding water part side mouth. Portus Rutulenis or Ratupie, Kent, the river side bank or ford part port or surrounding water part side. Portus Lemanis, Kent, the lower water place part, haven or port. Potheridge, Devon, the surrounding water side edge, ridge or promontory. Pottesfleet, York. the surrounding food water part. Potton, Bedf. the surrounding water part side town. Pouderham Castle, Devon, the surrounding water part vale caste. Pouderbach Castle, Salop, the small or brook surrounding water part castle. Hence Peach, Pock, Pocket, Poculent, Poke and Poker, the surrounding part or water part, action or shut; Pod, surrounding the part or side; Poem, on the surrounding parts; Poignant, Point, Poifon and Po se, on the surrounding part side or edge; and thence the point or edge of a weapon; Pole, the surrounding water place part; Police, Policy, Polish, Polite and Polity, the surrounding water place side lect or palace; Poll, the head or high surrounding water place; Pollute, using the surrounding water place; Polyglot and Polygon, on the surrounding water place or many tongues and angles; Pommel, a hand, the foot or other part on the surrounding water place; Pomp, on the surrounding part; Pond, a water part inclosed on the side; Ponder and Ponderous, inclosed or shut parts; Pontage, going over a border river or bridge; Pontiff and Pontop, on the surrounding water top or a floating bridge or ferry boat; Pool, an inclosed water place; Poop and Pop, the hind part; Pope, a general bishop or father; Populace, every country or people; Perch, the fore inclofure; Pork, the quill or brille surrounding or fat part; Pore, on the surrounding water part and port and its derivatives; Portion, on the surrounding water part side; Pore, the surrounding water part side or the head low or down; Poffes, the surrounding or inclofed side part; Post, the surrounding part or border lower side, after or behind; Pot, a side surrounding water part; Potent, and its other derivations, on the side surrounding water part; Pouch, an inclofed or shut part; Poultry,

Poulty, on the inclofing fpring or fowl tribe ; Pounce, a fur-rounding part inclofed ; Poundage, the inclofing water part or poundedge or pontage, the furrounding water part ; Pout, the part out ; Powder and Power, the furrounding fpring part.

PRESIDIUM or Warwick, War. Corfica, the border water part fide ftation, garrifon, out guard, prætenturæ or water part below at the fide of the country. Prætorium or Partrington, Yorkf. Scot. Italy, the inclofing water part or border town. Prætorium Metobriga, Spain, the fea water part confines præ-torium. Præfutagus, king of the Ifeeni and Bonduica's huf-band, the lower border fpring water part coaft chieftain or guardian of the firft or loweft river borders Pregaer, Corn. on the inclofing water part. Prendergaft or Clan Moris, Ire. on the inclofing water part fide or fea coaft. Prenvol or Bren-wall, S-Wales, the high or hill inclofing water place or vale. Frethut, Wilts, the inclofing or fhut water part. Preftan or Preftean, S-Wales, on the lower inclofing water part fide, ftation or præfidium. Preftholm. N-Wales. the lower water part fide ifland. Prefton, Lanc. Scot. the lower water part fide or ftation town or a præfidium. Prinkneith, Gloc. on the hill inclofing water fide Prittlewel, Eff. the water part at the fpring place or the priors dwelling from their fituation thereon. Procolitia, Protolitia or Prudhow, N-hum. the firft inclofing water place or river fide ftation. Prom-hill, Kent, the furrounding water part bank or hill. Pryfetes Flodan, Suff. the fpring water part on the flood part. Hence Practice, the fpring water part lower fide action or a river at the fea fide ; Precipe, the inclofing water confines part or keep ; Praife, a thing on the water fide or found ; Prank, on the water part confines ; Prate and Prattle, a thing at the water fide or place ; Proxity, the fpring water part fide or property ; Pray, a part fpringing upwards ; Preach, a praying action ; Prebend, the fpring end or upper fide part ; Precede, the fore fide ; Precept, the keeping part ; Precinct, the inclofing water part fide ; Precifion, the part above the low part and its deri-vatives ; Precith, over prifed ; Predeftine, before the part or fide, that is in ; Prefture, the fide or fore fide part Prefect, the firft on the water part fide ; Prefer, by the water part ; Pregnant, on the inclofing fpring water part confines ; Prelate, the water part place fide ; Preliminary, on the water part limits or edge ; Prelude, before the high fpring flood fide ; Premature, before its fpringing to the fide or exiftence ; Premier, the middle or divifion water part man and its derivatives ; Prentice, the lower or water part fide or in the houfe ; Preparation, the fore part at the fide ; Prepofition, putting before ; Prerogative, the

water

water fide border or circle part property; Prefage, fore faying an action; Prefbyter, a prieft, or by the water part fide one; Prefcription, on the water part border fide or country; Prefent, fent before; Preferve, to keep; Prefide, the water part fide or gar ifon; Prefs, lowering or leffening the water part; Pretence, the part below the fide or right; Pretor, the water part border man, a judge or ruler, Pretty, at the water part fide; Prevail, the fpring water part flowing; Prevaricate, fhutting the fpring water part, Prevent, the fide on the fpring water part; Prey, on the water part fide; Price or Prize, the water part fide, where things we e apprailed and fold; Pride, the firft divifion of property; Pride-gavel or Lamprid, the furrounding water part fide or holding; Prim, Prime and Primate, the firft or water part rim or border, and its derivatives; Prince, on the inclofing water chieftain, with its derivatives; Print, a thing on the fide; Prior, the circle water part; Prifm, the fide furrounding fire part or fun below; Prifon, on the fide part; Prithee, pray thee; Private, privy at, or at the fpring water part and their derivations; Probe, Probate and Proof, the furrounding water part and their various derivatives; Procefs and Proclaim, on the inclofing water fide; Proctor and Procurer, the inclofing fpring water part or deputy actor; Prodigal, the furrounding water part of the Gaulifh or alley fide; Prodition or treafon, on the furrounding fpring water lower fide; Profane, the part in the furrounding water part; Profefs, the part at the fide of the furrounding water part, which muft be through every part of the country; Profit, the fide furrounding water part; Profligate, flying the furrounding water part gate; Profound, the fide or country inclofing fpring water part; Profufe, the inclofing fpring water part ufe; Prog, the furrounding water part confines; Progeny, on the inclofing water part; Progrefs, the furrounding water part to the lower water or border; Prohibit, the furrounding water high part; Prolix and Prolong, the length and lengthening, the furrounding water part; Prominent, on the fide or edge of the furrounding water part; Promife, for the meres or mark; Promote, for or over the furrounding water fide; Prone, within the furrounding water part; Prop, the furrounding water part up; Propagate, acting at the furrounding water part; Propel, for the water place part; Propend and Propenfe, the furrounding water part up end on infide; Property, by the furrounding water part fide; Prophet, the furrounding water part fide part or perfon; Propinquity and Proximity, on the furrounding water part border or fide; Propitious, the furrounding water part fide part; Propofe or Propofite, fetting the furrounding water

water part ditch; Propt, the surrounding water place side up; Proserine, below the surrounding water border part; Prose, the surrounding water part, side, as verse is the spring water side; Prosecute, dividing the surrounding spring water part, as sue or issue is springing from below; Prospect, the side water part below the surrounding water part; Prosper, by the surrounding water part side; Prostitute, the surrounding water part side at the side or towards; Protect, covering the surrounding water part; Protest, the surrounding water part side; Prothonotary, the surrounding water part first notary or register; Protract, the surrounding water part prolong; Proud, the surrounding water part springing or flowing part; Protrude, the surrounding water part dwelling part; Province, on the surrounding water part side or edge; Provoke, the surrounding water part mock; Provost, the surrounding water part next or chieftain; Proxy, for the chief; Prude and Prudent, from rude; Prune, from springing; Prurient, a thing being on the spring or itching

PUCKERIDGE or Puckerich, Herts, the spring water upper or ford part. Pucklechurch, Glo'c. the spring water part place church. Pudeford, Devon, the spring side way or ford. Pudsey, N-hum. the spring part at the sea. Pudeln, Oxon. the spring part bank. Pulton, Wilts, a town on the spring part. Purbeck, Dorset, the spring water part brooke. Purford or Pyrford, Surry, the spring water part way or ford on the river Wey side. Pusey, Berks, the spring part side. Pushall, Oxon, Dur. the spring part side hall. Puteoli, Italy, the spring part at the water place or sea side. Putney, Surry, on the side spring part. Hence Puberty, beside the spring part. Pubescent, the growing spring part; Public or Public, every inclosing spring place; Pudder, the spring side part; Pudding, the inclosing water side spring part; Puddle, a water place at the spring part; Puff, the spring part on; Puny, on the spring part; Punish, on the spring side; Punk, on the inclosing spring water part confines; Pun, on the spring part; Pup, spring up; Puppet, a springing up thing; Purchase, the spring water part inclosure; Pure, on the spring part; Purge, the spring water part edge or action; Purle, the spring water place; Purlieu, over the spring place; Purlin, the spring part or covering lines; Purple, the spring light part; Purport, the opening of a discourse or spring part; Purr, on the spring part; Purse, the spring water part side; Pursue, the spring water part issue; Purtenance, on the side below the spring water part; Pus, a thing springing from below; Push, Put, Pustule and Put, the spring part at or to, as to the sea or any other inclosed water

U   place,

place, either animal or merely earthen; Putative, Putty and
Puzzle, the spring part low; Pyre, the spring water or fire
part, or in a secondary sense, a long part; Py amid, the boundary pyres, or the flame of fire terminating in a point, hiero-
glyphically denoting the limits of a kingdom, and registering its
antiquities, as well as serving for cemeteries of the priests
and princes of the sun, and denominated hieroglyphics from the
secret or sacred scriptures thereon; but those on the British
columns and pillars were probably erased by the Romans or
effaced by time: for the Druids, as well as all other Pythago-
reans, seem by their language and doctrines to have been once
possessed of the mysteries of the sacred characters, as appears
by those of Apis and the Egyptian pyramids, the Paphian
Venus, Bal Peor or Priapus of the Cananites, the Jaggernaut
Blackstone or god Brama of the Gentoos, and the Venus U-
rania of the Persians.

RABY, Dur. an habitation on the water or river part or the
abby. Rabble, the country place or race. Rabe, Hungary,
the river part. Rabbet, the ground springing part.

RACLINE or Ricinia, an island of Scot. part in, the inclosing water place or surrounded by water. Hence Race, as
proceeding from, issuing or flowing like the spring water part;
Rack, the water part shut; Rackrent, the water part shut or
border rent; Rake, the gathering part.

RADBURNE, Der. on the spring water or brooke side part
or country. Radcliff, Salop, the cliff side part or country.
Radmild, the middle or boundary side part or country. Rad-
nor, S-Wales, on the border side part or country. Radwinter,
Ess. the country at the spring rising part. Hence Radiant, on
the water part side, or in the other language, the fire division
or particles on the lower parts; Radish and Radix, the lower
side or ground part; Raft, the rope or water part tie.

RAGLAN or Ragland, S-Wales, the water part bank, bank
side or land. Rage, Nott. the inclosing water part. Hence
Rag and Paget, the divided parts or a firey or hot action;
Ragout, an hot gusto.

RAGS, Ess. the surrounding water part or a country on the
water part. Rhadr Gwy, S-Wales, the Wye or spring high
or roaring water part. Ralegh or Raleigh, Corn. Devon, the
inclosing water place. Hence Rail, the division or inclosed
place; Raiment, the cover in a part; Rain, the inclosing or
high water part; Rally, the water part from below; Rainy and
Rallery, on the water place.

RAMSBURY,

RAMSBURY, Wilts, the furrounding water part fide burrow. Ramſden, Oxon, on the furrounding water part fide or vale. Ramſey, Hunt. Scot. S-Wales, the furrounding water or ſea part. Ramſgate, Kent, the furrounding ſea part gate. Ramſgrave, the furrounding water part warden. Hence Ram, the covering part; Ramble, the furrounding water part place; Ramme, the furrounding water part ram; Ramiſh, Ramp and Rampant, the furrounding or covering part.

RANDALS Trenches, Scot. the incloſing water part vale trenches. Randavan or Ranfrew, Scot. the incloſing water part or ancient neighbourhood. Hence Ran, the incloſing water parts; Ranch, Rancid, Rank, Rancle and Rancour, the incloſing water part fide or border; Rand, the incloſing water part fide; Range, the incloſing water part confines; Ranſome, the incloſing water part or confinent turn; Rant, confinement part or property; Raſe, the part over the lower; Raſcal, the female cull

RATÆ, Ragœ or Rothby, Leic. the water part coaſt, confines or habitation. Rathcomire, Ire. the water part fide vale. Rathdown, Ire. the water part fide down, plain or low part. Ratley, Radley or Rotelei, War. the water part fide place or palace. Ratoſtibius, S-Wales, the water part border or fords below the ſeivi fide. Rata, Scot. the fide or country water part. Hence Rat, Rate and Rather, the water fide part; Ratio and Rational, accountable, and Rattle, as appertaining to the fide water parts.

RAVENGLAS, Cumb. the river fide place, or in a ſecondary ſenſe, the river fide green part. Ravenſbourn, Kent, the river fide brooke. Ravenſere, York. on the river fide. Ravenſwath, York. the river fide water. Ravius, Scot. the fide ſpring water part. Raunten, Leic. a town on the ſpring water part. Rawdon, York. on the ſpring water fide or town. Rawdikes, the ſpring water dikes. Reyland, Gloc. the ſpring water part bank fide or high land. Hence Ravage, the ſpring edge or flowing action; Rave, the ſpring or flow; Raven, the ſpringing or ravenous one; Ravenous and Raviſh ſpringing the female or the female ſpring; Raw, on the ſpring or juicy part; Ray, a fire part or particle; Razor and Razure, the upper part from the lower.

REA, Rhea, Rhy or Rian, Berkſ. Yorkſ. N-hum. Wales, Ire. the water or flowing water part or one. Reading, Berks, on the incloſing water part or river fide. Readliquite, N-hum. the river fide borders. Reamn, the water vane. Hence Reach, the river confines; Read, the river or flowing diviſions or parts; Ready, at the river fide; Reck, on the river or water

place

place or a county; Realm, the surrounding river or water place; Rear, on the river or flowing or fishing water; Reason, a reckoning, or on the river side, where registers of accounts and government records were kept, as at the courts leet and baron; Reciprocal, the ebbing of water or the reverse of its fluxion or; Rel as and Rel, respecting Belinus the prince or river warden; Rebound, Rebuff, Rebuild, Rebuke or a return of them.

Rechii, Yorkf. the cell or hidden river. Rech Diche. Camb. the water part ditch or river. Recine, an island of Scotland, in the water part. Recolve, Racculf or Regulbium, Kent, the helm or spring water place. Hence Receive, Recent, Recci, Reckon, Recine, Recluse, Record, Rectify and Redeem, as qualities and properties appertaining to the boundary river or water part, which have been already explained.

Redburn, Herts, the water part or river's neighbourhood. Redbridge, Hants, the river side bridge. Redditch, Soc. the river side inclosure. Redeham, Norf. the water part or river side vale, home or dwelling place. Redhead, Scot. the river or water part side head or promontory. Redwell, Ess. the water part side spring or dwelling place. Redmyre, Yorkf. the river at the sea part. Redbrooks or Dalreudin, Scot. the river, water part or spring side vale or on the river side or vale holders. Hence Red, or rather, from it-ed, the fire part; Reddition, on the water part side or the return of the fire; Redolent, at the spring or river side bank or vale; Reduce, lessen the the spring water part; Redundant, the flowing at the spring water side; Reed, the river side; Reeve, the river part warden; Refuse, in the river part; Reflect, the flowing water place side; Refluent, a flowing back of the water; Refresh, the river part side; Refuge, the spring water part confines; Refute, below or aside the spring water part; Refute, without the flowing or rush.

Regia, Regimm, Rhegmm, Reginium, Rhine or Rhing, Rarin, Germany, Elysium, Italy and Wales, the inclosing ditch or boundary water confines or on the same. Reinfac, Scot. the inclosing water part neighbourhood. Reingrave, the inclosing water part warden. Hence, in, within the inclosing water part; Regal, on the boundary water confines or side or river; Regard, the water part guard; Regency, regia, regium and regilder, on the inclosing water confines or side; Regilder, on the inclosing confines; Regrate, at the inclosing water part where things were sold and forestalled; Regress, below the inclosing water part; Regular and reign on the inclosing water place, palace or kingdom; Reins, as having
the

the property of separating the urine from the blood, like the production of rain from the waters below; Rejoin, the rain water returning to their former body; rejoice, the sun's light or fire below; Reapse, below the spring water place; Relate, on the water part side; Release and Relinquish, below the inclosing water part place or bank; Reticks, the flowing water part confines, where things are left at the ebb; Relieve, the flood water part; Relish, the flood water part side; Religion, on the flood water place confines, the usual places of religious worship; Rely, on the water part.

REMSY or Romney, Sussex, on the surrounding water part. Remi or Reims, of Belgic Gaul, the surrounding water part. Hence Remain and Remains, on the surrounding water part, or the relicks of the sea; Remark, the water part mark; Remedy, the mode or middle water part; Remote, the surrounding water part side. Remove the surrounding spring water part; Remount, on the side of the surrounding spring water part.

RENDCOMBE, Glouc. the inclosing water part side vale. Renlisham, Suf. the inclosing water part side lover; the c palace, leet, or ancient vale or dwelling place. Renes, Brittany, the inclosing, or on the water part side. Hence Renard, on the water part side governor, or a fox in a secondary sense as possessing the properties or qualities of a governor, and also the virtues of too many of them according to the old English adage of setting the fox to keep the geese; Rencounter, on the inclosing water part together; Render the inclosing water part which divides the country; Rendition, the inside of the inclosing water part; Renegade, leaving the inclosing water part; Renew, the water part in a spring; Rennet, the water part in the inside; Rent, the inclosing water part division or portion.

REPENDUNUM or Repton Der. a town on the spring water part side or end. Reppen in Pomerania, the water part end. Reogh, or Wexford, Ire. the water part confines or the spring side way or ford. Hence Repair, the spring water part pair or sides; Repass, the water part pass or side port; Repay, the water part part; Repeal and Repel, drive back the water part; Repeat, the water part at; Repent, the water part end or head on the side; Repine, the spring water part in; Replenish and Replete, the spring water part in the water place sides; Replevin, the water place edge; Report, the water part port, or gate; Repose, the surrounding water part side part; Represent, the water part present; Repress, below the water part; Reprimand, the water part prince's command; Reprisal, the price on the water part; Reproach, returning from the inclosing water part; Reprobate, rebate or blunt, the surrounding water part

part; Reptile, the water part fide place, or race; Republic, the incloſing water part place, or nation; Repudiate, without or from the fide ſpring water part; Repugnant, putting back the ſpring part; Repulſe, backing the ſpring part aſide; Repute at the ſpring water part; Requeſt, the incloſing ſpring water part lower fide; Requite, the ſpring water part fide.

RERECROSS, Yorkſ. W-mor. the water part or river border fide or croſs, which was the croſſing and market place. Reretby, York. on the water part fide habitation. Rerigonium or Carrick, Scot. on the incloſing water part or upper border fide. Hence Rear and Rereward, on the water and ſpring water fide or border.

RISANIA, Ruſſia, on the water part fide. Refiſtum Thrace, the water part lower fide. Reſovia Ruſſia, the water part or river fide or way. Hence Reſeve, the water part fide ſhut; Reſent, ſent back; Reſerve, keep back; Reſide, the water part fide; Reſt, at the water fide part; Reſtore, to the water fide border; Reſult, the water part at the ſpring water place, which flies back; Retail, the water part coaſt place; Retain, on the water part coaſt; Retard, on the water part coaſt fide; Retire and Retreat, the water part fide; Retort and Return, the water part or fide turn.

REUDINI and Rendigni, Scot. Germany, on the ſpring fide part. Reveſio or Ruelio, Gaul, the ſpring water fide. Hence Reve, the river fide Warden; Revel and Reveal on the ſpring; Revenge, the incloſing ſpring water confines; Reverb, ſpringing back; Revere, on the ſpring; Revert and Reverſion, a part or thing ſpringing back; Review, the fight ſpringing back; Reward, on the ſpring water fide or the warden's return; Reveland, the river bank fide or land, which at firſt were the Thane land

RHEBA, Ire. the river part. Rhedeſdale, N hum. the river fide lower vale. Rhidol, Rhidale, or Ridal, Yorkſ. W-mor. Wales, on the river fide or the river Vale. Rhiegate, Surry, the river gate. Rhos or Ros, Heref. Scot. Wales, the ſurrounding water part fide or wet boggy ground uſually on the borders of countries. Rhobogdinm, Ire. the ſurrounding river water part fide. Rhydland or Rutland, But. N Wales, the river fide, or bank fide or poſſeſſions. Rhutupiæ, Richburrow or Sandwich Kent, the river fide way or ford part, the upper ſpring part or the ſand ſpring water confines. Rhibel, Ion the river water place. Hence Rhapſody, a flowing from the found diviſion; Rhetoric, the border or mouth flowing action; Rheum, the ſurrounding ſpring; Rhom, the ſurrounding water part or four fides; Rhyme, the ſurrounding ſpring.

RIBOD-

RIBODUNUM, or Ribchester, Lanc. a town or city on the surrounding river water part. Ribel, Lanc. the division or river water place. Ribston, Yorks. the river or division water side town. Hence Rib, the division part; Ribbon the rib one.

RICAL, Yorks. on the river or division water. Richmond, Surry, Yorks. a mount or hill on the river or division water confines. Rickmansworth, Herts. the division or upper in-inclosing water side spring part. Ricot, Oxon. the river or division water coast. Hence Rich, the division or inclosing water confines, a kingdom, riches, or the uppermost part; Rick; the uppermost.

RIDAL, W-mor. the division water or river vale. Riddington, Oxon. the inclosing river side town. Ridley, Chef. the river or division water side or ford place. Riduna or Alderney, on the division or high water or ford part. Rid, and Ridin, a ford or paffage.

Riga, and Riga Altera, the division water confines and the division water confines other side. Rigodunum or Riblechester, Yorks. a town or city at the river or division water confines. Hence Rig and Rigging, the division water confines; Right, the division water confines side; Rigid and Rigour, the inclosing water side. Rill, the flowing water place or brook; Rime, Rims, and Rimy, the river border.

RINGWOOD, Hants. the inclosing water side or division wood. Rin-rom, be. between the flowing water. Rinny, Wales, the melting or division water or river. Hence Rind and Rinsing, the inclosing part; Rink, on the river side; Rib, division of the surrounding water side.

Ripley, Surry, boiling up the division water or river place. Ripon, Yorks. the division water part. Hence Rip, a division, Rips, a part up; Ripple, the river or washing water place.

Rippley, Derb. Yorks. the flowing water place spring or river side or borders. Kitchin, Yorks. on the inclosing river side. Riddun and Riddin, W-mor. N-ham. a vale on the inclosing river side. Ritingho. Bed. on the high river confines side. Hence Rife, ribble, rich.

Rirox, Yorks. on the division water or river side, or the ford or water paffage. Ribromagus, Gaul, the surrounding water ford. Hence Rion and Ritual, on the river side, usual places of religious and other ceremonies among the Gaulish druids, who prefided over the fide rivers, which introduced the opinion of the nymphs and the Britons, paying adoration to the springs and rivers, though it seems much more probably from our deduction of names, that the Romans were the first

adorers

²dorers thereof and of British men and women or rather colonies, as Minerva, Hercules and others, appear to be Phrygian colonies, but Roman deities, and the inscriptions on stones
produced in support of this mistaken notion, are all Roman.

ROADGATE, York, the surrounding water part side or roadgate or way. Hence Roach, Road, Roam, Rem, Rear, Roa-
ry, as properties of the surrounding water or river.

ROBOGDII, Ire. the surrounding water part side or the water
bogs. Hence Rob, Robbery, Robe, and Roburt, as incident
to the surrounding water side.

ROCH and Rochdale, Lanc. the surrounding water side or
vale. Rochester and Rochford, Kent, Ess. the surrounding
water side, city, ford, or way. Rockley, Wilts. the surrounding water or rock place. Hence Rock and Rocky, as moving
like a boat on the water and rocks, and stones are mostly at the
water side.

RODAM or Rodan, Salop, N. thumb. the surrounding water
side. Roding, Ess. the inclosing water side and perhaps in a
secondary sense the Ingles surrounding water side. Rodleigh
or Rodway, near the surrounding water side place or way.
Hence Rod, as growing on the river side; Rode, as reeds
were along the river sides; Rogation, on the surrounding
water side.

ROLL-RICH -stones, Oxon. the surrounding water place,
country, or kingdom roll, round, limit, or relise, of
which there were many, as those called maini, oyniled or
the maine ary or reulter stones, and rolmen, hirien, or the
long boundary stones in Pembrokeshire, cryg y vyan or the
kingdom's stones, and buarth Arthur and maen gwyr, or
Arthur's pinfold, or spring water barlie stones, in Carmarthenshire, hir vaen gwyddog, or the long gwydhic water confines
stone between Carmarthenshire and Cardiganshire, cader idris
or the lower confines station in Merionethshire, and eeise gui-
thiel, or the spring water division or top roman stones, in Anglesey. Hence Roll and Rolland, a place or country included,
inrolled, fenced and bounded.

ROME and Romen, Italy, on the surrounding water part.
Romara and Romney, Som. Suss. on the surrounding water
part. Hence Room, Romage, and Rummes. The Romans,
in order to accustom the provinces to their religion, laws,
customs and manners, placed colonies of romans with their
legions, as that with the ninth legion at Camalodunum,
which Boadicea destroyed, and many of the lost bands settled
amongst the Coritani, and that at their municipium or free

city

city Verulam, but the prefidiums were only garrifons or frontier towns as Warwick.

RONA, one of the weſtern iſles of Scot. a part or country
in the furrounding water or fea. Hence Ront or Round, Roof,
Rooſt, Root and rope, the furrounding incloſing or incloſed
part ; Roration, the furrounding water.

Ros or Cantæ, S-Wales, Ire. Scot. the furrounding water fide
or coaſt. Roſſal or Ruſſel, Salop, on the furrounding water or
river fide. Roſearock and Roſeland, Corn. the furrounding water
fide rock land or ſtones, or in a fecondary fenfe, a wet ſtony
ground or moraſs. Roſcomon. Ire. the furrounding water fide
part or common. Hence Roſe and Roſet, or rather from the
colour and roundeſs of the fun.

ROTHER, Kent, Suff. the furrounding water part. Ro-
theram, Yorkf. the furrounding water fide or vale. Rotheſia
or Bute, Scot. the furrounding water fide lower part or iſland.
Rothes and Rothſay caſtle, Scot. the furrounding water fide or
iſland caſtle Rothwell, Route, and Routon or Rutupium,
Yorkf. Scot. Salop, the furrounding water fide well, or town.
Hence Rot, which is the effect of the furrounding wet ; Rote
and Rotund. a round part.

ROUSE, the furrounding fpring fide. Rouſcliffe, Cumb. the
furrounding fpring fide cliff. Rowte, Ire. the furrounding water fide. Hence Round, Rouſe, Route, Row, Royal, on the
kingdom or furrounding water part.

RUCHAL, Ed on the incloſing water place. Rud or Rut,
and Rudheath, Rut. Chef Wales, the river fide or heath.
Rugby, War. the fpring water confines part or habitation ;
Rongemont, the fpring water confines mount or hill. Rules,
Wales, the fpring water place fide. Rumley and Rumford,
Yorkf. Ed the furrounding water place, ford, or way. Run-
corn, Chef. the incloſing high fpring water. Rune. Ire. the
fpring water part. Ruſhbrooke and Ruſhton, Suff. N-ham. the
fpring fide brook or town. Ruthin, N-Wales, on the fpring
fide. Rye and Ryal, Suff. Yorkf. Rut. on the fpring water
part. Hence Ruſh, the fpring part ; Ructation. the fpring
water coaſt ; Rude, the fpring fide place ; Rue from the fpring ;
Ruſ, the fpring confines or covering ; Ruin. in the fpring ;
Rule, or on the fpring water place ; Rum and Rumer, the incloſing fpring water ; Run. in the fpring ; Runt, in the river
part ; Runtune, the fpring part fide ; Ruſh and Ruſſ, the
fpring fide ; Rut, the fpring part.

SABRIDGEWORTH, Herts. by the lower fide fpring water
bridge. Sabrina or Severn, Glouc. the lower and

diviſion ſpring water part or river. Sabine, Italy, the lower end or incloſing part. Hence Sabbath, the lower or reſt part diviſion; Sable, the lower or ſhut place; Sabulous, the lower or ſide ſpring water place; ſ being moſtly a derivative of c, and its moſt primitive meaning may appear from the radicals deſined under that letter; as for inſtance, ſide a compound of ſi de, the lower diviſion or part or the part where the river meets the ſea, ſeems to derive its origin from the Welſh cid, together or from ci de, the diviſion and incloſing water together and the radical of k, as kiſs, the river or ſpring and the ſea meeting together.

SACY Foreſt, N-ham. the lower water confines foreſt. Sackville, Suff. France, the lower incloſing water dwelling place. Sacole, Spain, the lower incloſing water place. Hence Sack, incloſing the lower or the lower incloſing water confines; Sacred, the lower incloſing water ſide, which was deemed ſacred from its being ſecret and the uſual places of worſhip or cells; Sacriledge, robbing ſacred or ſecret places; Sage, Sake and Saker, as belonging to the lower water confines; Saint, on the lower or ſea ſide or the cell; Sail, on the ſea.

SADURNIA, Saturnia or Italy, on the lower water part. Saturium, Italy, the lower water ſide. Hence Sad, the lower part; Saddle, the lower or below the ſide place.

SAFFRON, Eſſex, on the lower or ſide water part. Hence Safe, the lower or ſide part; Saffron, from the name of its place of growth.

SALENE, Salinæ, Salnedy or Calene, Bedf. the ſide lower or incloſing water banks, cells or groves, as thoſe of the Silures, Selwood &c. Saltry or Saltion, Scot. Camb. the ſide or lower place town. Salwarp or Saltwood, Kent, Wore. the low or ſalt ſpring water part, ſide or wood. Hence Salacious, the low or female action; Salad, at the water place ſide; Sale, the ſide or lower water place, where the markets were held for the ſale of all ſorts of cattle, goods, and merchandizes, and the toll or the boundary part paid; Salient and Saily, the ſide or low water place; Salmon, in the ſurrounding lower or ſea water place; Salt, the lower water place or ſea ſide or part; Salt, the lower ſpring part; Salute, the lower ſpring ſide.

SAMOBRIVA, in Gaul, the lower ſurrounding or more water part. Samos and Samen, Greece, the lower iſland or ſea ſurrounding parts. Samnium, Italy, on the ſurrounding ſea lower water part. Hence Same, the ſurrounding water place which are the ſame; Samphire, on the ſea banks parts;

Samoby,

Sample, the same place or the two banks or sides of the surrounding water place.

SANBRIOC, in Gaul, on the lower sea water part. Sandale castle, Yorkf. the lower castle in the vale. Sandgate, Kent, on the sea side gate. Sandon, Staff. a town on the side or lower water or spring. Sandihead, Scot. a head on the sea side. Sandwich, Kent, on the sea side, street, shut, fortification or city. Sanghar, Scot. a town on the lower water confines. Santon, Devon. a town on the sea side. Hence Sancible, enabling the lower; Sanction, Sanctify, and Sand, on the sea side or bank; Sandal, upon the lower part; Sane, the lower up; Sanhedrim, the holy over or along the rim or borders; Sank, low in the water.

SAPCOT and Saply, Hunt. the lower inclosing water part side and the lower place. Saperton, Gloc. the lower or side water part town. Sapis, Italy, the lower side water part. Hence Sap, the lower part; Sapient, the inside part; Sapor and Sappy, or juicey.

SARAGOSA, in Spain, the lower country or water confines. Sarden, Staff. the lower country or water side, den, or vale. Sarrifberia, Sarum, Saltbery, Sarnbirig, or Sorbiodunum, Wilts. Lanc. the lower country or side water place part, abode, burrow, or town. Sarn halen, N-Wales, on the lower water part bank, or the paved water bank or causeway. Sarmatia, the lower country or surrounding water part side. Sarnia or Guernsey, in the inclosing water lower part or island, or the lower water causeway, or streets over which probably the first Britons passed into this island. Saronide, on the lower inclosing or sea water parts, as Druids were on the spring water parts or river sides. Hence Sarcotick, the outer coat or covering; Sark, a shirt, the lower covering; Safa, shutting the lower or inclosing.

SATAEREGIA, Dur. the lower or sea water part confines. Satcula and Satio, Italy, the lower side inclosing water place. Hence Sat, the lower side or part; Satan, in the lower coast or the lower below; Satchel, inclosing the lower part; Satellites, the lower or lesser covering or sky lights; Satiate, at the lower or inside side; Satin, on the lower part; Satisfy, the part below the spring part; Saturate, at the lower or somuse water parts; Saturday, the lower water part or sea civil use; Saturn, on the lower water part or sea side; Satyr, on the spring side or woods.

SEVARN, Severn, or Havren. Salop. &c. the side or lower inclosing spring water part. Sevenby, Wor. on the lower spring part or abode. Savernake forest, a forest on the lower ſpring water. Savil, the lower ſpring or dwelling water place. Sauldon, Bucks. a den or vale below the ſpring water place. Sauhguer caſtle, Scot. the lower border water caſtle. Hence Savage, below the incloſing water confines dwellers or ſeamen. Savanna, on the ſpring water ſides or lower parts; Sauce, Saucey, Savour, and Sauſages, the lower or female ſpring water confines; Sawpit, the lower ſpring water part; Save, the lower or ſide part

Saxham or Saxmundham, Suff. the lower water ſea confines or Iſceni, vale or home. Sax or Sex, Spain, the lower water confines. Saxani France, the lower water confines or coaſt. Saxones, Saeſons, Saeçons, Sacen, Axones. Holt-Saxons or Foſi of Spain, Gaul, Britain, Ire. Denmark and Germany, on the lower or weſtern water or ſea confines, the high water place ſide Saxons, and the ſurrounding water part or ſea ſide, or in a remoter ſenſe the ſons or inhabitants of the ſea confines. This ſeems to be an appelative made uſe of by Pliny, Ptolemy. and other Roman writers to deſcribe the Ventalſecnorum and other Britains, who had been diſpoſſeſſed by the Roman colony and ninth legion ſettled at Camalodunum of their country, and taken poſſeſſion of ſeveral parts on the ſea coaſts of Ireland, Scotland, England, and Germany, and made depredations on the Roman provinces according to the Britiſh poets, and from which the Iriſh Scots ſeem to be deſcended, but meant no particular nation till after the Romans quitted Britain and Germany, and as X and Sax are no German, Saxon or other Celtic radicals, but are of Greek origin, as appears by a paſſage in Homer, which ſays that the Greek name of the Phrygian river Scamander was Xanthus, and enter into the compoſition of theſe dialects merely as derivatives from the Latin, the name Saxones muſt be of a Roman origin, and the rather becauſe in the Saxon dialect, it would have been wrote Seecens or Secons according to the following Saxon names, viz. Hend, Concani or Cantabrians, Secen or Segontine, Can. Gofen or Connel, the liberty of the Sole or Saxon, Secca or Sea and Trigontes or Britiſh Saxons, Can. Canton, Kent, Thorn Token, not Saxon, Aiken and ſet and Lot, or the ſea ſide and river border toll. or Scots and Engliſh toll, and ſo. The Saxon dialect, though in the concrete, it differs in word and form the other Celtic dialects, yet in the abſtract varies little from them, particularly the Welſh and Engliſh. The Saxons who arrived in Britain in the aid of Vortigern or to the ſeveral countries where they had been diſpoſſeſſed ſome, reſtricted to themſelves the original names of ſeveral parts, as appears by their early writing, and their being the models on which the English and other British dialects were framed, which are rarely to be met with in any other parts or countries. And as to the language and antiquities of Bede and other monks, they as little correspond with the true dialect and affairs of Saxon Britain, as the conceits and absurdities of some others, with respect to a regular succession of Trojan princes in Britain, at a time when the island was under the government of the Druids in temporalities, as well as spiritualities; nor are the pious and angry declamations of Gildas, another wise leader of our antiquities on the vices of his countrymen, and the barbarities of the Saxons, Picts, and Scots, to be regarded as exact representations of the transactions of those times; neither can it be believed that those people, who are said to have fought a battle in the north against the Scots and the Picts in aid of Vortigern, whose names of persons and places are mostly to be met with in Scotland and Ireland, were any other than the Scots and Picts themselves, and the Irish and some Northumbrian Britains; for Gildas and Bede, seem to consider the Scots, Picts, and Saxons, as joint invaders of the Britains, and the circumstances of that affair are very differently related and incredible in themselves; and those very Saxons, immediately after this pretended decisive battle against all the northern nations in favour of the Britains, soon after appear to have entered into a league with the Scots, Picts, and Northumbrians, against them, under Octa the brother and homager of Hengist, duke of Kent; and ever after living peaceably together with the assistance of some of the other expelled Britains from the continent, pursued their conquest of Mercia, East-Angles, Kent, and other countries of the southern Britains. If we seek amongst the historians for the Saxon origin, Ammianus Marcellinus will inform us, that they were those who invaded Britain in the days of Valentinian, and that they and the Cumbri came out of Greece with Hercules. Caesar says, that there was an old tradition, that the Brigantes, a mixture of Saxons and Britains, were the Aborigines of Britain; Crantius says that they came from the German Catti; Whidichind, that those who came to the aid of Vortigern had no fixed home; Ptolemy, who first mentions them says, that they were from the Chersonese Cumbri or Denmark; Ethelward, that they inhabited a country between the Rhine and Denmark; Janus Douza, that Hengist came from Batavia or Holland; Procopius, that the inhabitants of Britain were the Angles, Frissons, and Britains; Ubbo Emmius, that the followers of Hengist were mostly Frissans; Claudian, in his commendation of Valentinian, seems to place

the

the Saxons, who invaded the Southern parts of Britain in the
Orkneys, and the Scots as coming from Ireland, and with
which the Welsh antiquaries seem to agree; Egesippus and
Isidore say, that the Saxons had their name from their situation on the sea shore; Gorogius Turonensis, that they came
from Baieux or lower Normandy; Eginhardus, that the Saxons of Germany, were descended from those of Britain; Pez-
ron and Verstegan, that the Germans, Gauls, Welsh, and
Britains were the same people; Camden, that the Saxons
were from the Saxones of Scythia; Annius Viterbienses, and
Suffredus Petrus, that the Saxons and Cimbri were the remains
of Alexander's army or Greeks, which honest Sherringham
doubted, as he did not think they spoke Greek, but many of
their vocables, and names of places appear to be Greek; Tri-
themius, out of Wastaldus says, that the Saxons were the offspring of the Cimbri and Trojans; Lazius and Capnio, that
they and the Cimbri were from Phrygia, and that their language was the Phrygian; Sir William Temple, that the customs and manners of the Physians or Phrygians were English; Bede, that Augustin took some Franks from Gaul to be
his interpreters here, that the Saxons took possession of the
Orcades and northern parts of Scotland, and joined the Picts
against the Britains, thus looking upon them as I do, to be
the same as the Scots; that Wilfrid, Suidbert, Willibord, and
other Englishmen who preached the gospel in Friezeland spoke the
Frizian tongue as if they had been natives; Paul the deacon, Bede,
Gildas, Boniface, Widiscind, and others, suppose that the Britons,
Angles, and Frizons to be the same people, and that the Angles took their name from an angle or corner of the sea shore, which
they possessed, but that the Britains came to this island before
the Anglo Saxons; and others bring the Saxons from the neighbourhood of Brittany and Aquitain, that they were pirates,
and that they did not destroy, but incorporated with the Britains, and as Tacitus takes no notice of them, they were not
probably settled in Germany in his time. And however con-
tradictory those accounts may seem to others, they appear to
me to be pretty well grounded on facts, except Mr. Camden's
Scythian origin founded on the likeness of the names; since it
appears that some of every branch of the Japhetan race,
except some of Magog, Mesheck, and Tubal, have possessed some part of every country in Europe, at least on this
side Scythia, and the present names of places, which appear
to be the models of the English language, and the British
names imposed on the natural divisions of countries by its first
possessors before any towns or cities were built, though only
such

fuch of them as are mentioned in the Roman writings are fuppofed to be of Britifh original, and the reft Saxon, but the
Saxons being Britains themfelves, and fpeaking a Britifh dialect could only vary the names, by adding Ceafter, Burgh,
and other terminations to the Britifh radicals.

SCALBY Caftle, Cumb. at the fide or below the inclofing
water place. Scammey, an ifland of S Wales, in the lower
furrounding water place confines. Scarba, one of the weftern
ifles of Scotland, the lower inclofing water part. Scarborough,
Yorkf. the inclofing water fide or rock burrow. Scarrs and
Scarfdale, Der. the inclofing water fide rocks or vale. Sceaft
and Sceaftbury, Dorfet. the lower inclofing water fide part
or burrow. Sceorftan, Oxon. on the lower border or fhire fide,
or a boundary ftone in a feconary fenfe. Sceilcefter, N-hum.
the lower inclofing water place fide or city. Sci.edun, Devon. on
the lower inclofing water or border fpring, or on the fhire
fpring part. Selover manor, Suff. below, or at the fide of the
inclofing fpring water manor, Scordium, Cam. the ifland or
lower border, or inclofing water fide. Scotaney, Lin. below,
or at the fide of the lower furrounding water, or on the fea
coaft. Scots hall, Kent, the lower furrounding water fide,
place, or hill. Scotia, Greece, the lower inclofing water or
fea fide. Scotland, the lower furrounding water or fea fide
country or iflands, and in a fecondary fenfe, the weftern water
bank or woodfide or country. Scratchbury or Scracbury, Wilts.
a burrow at the inclofing water fide. Screeven, Yorkf. on the inclofing fpring water fide. Scrivelby, Linc. the inclofing fpring
water parcel le, part, or habitation. Scroby and Scrope, Nott. the
furrounding water part fide or border part or habitation. Scilly
iflands, in the Englifh channel, the lower inclofing water place
part, or iflands. Scythia, the lower divifion or inclofing water
fide, country, or poffeffions. Hence Scab, Scar, Scabbard and
Scabby, as covering the lower part ; Scaffold, Scalade, Scald,
Scale and Scollop, as being at the inclofing water place fide ;
Scalp and Scaly, the lower part ; Scammony, covering round
the lower part, Scantier, the leffer inclofing water part ; Scan,
the leffer fide or minute act on or inclofing water part ; Scan-
dal, on the fide of the inclofing water part ; Scant, below an hundred or the [...] Scare, the lower caps or inclofing
the water part ; Scarce, below the inclofing water fide ; Scare, on
border or inclofing water fide ; Scarf, covering the lower part ;
Scarp, the lower inclofing water part ; Scavenger the man on
the lower water part ; Sconce [...] or teen ; Sceptre,
the country or poffeffion [...] Schedule, the fpring
water place fign or [...] were manured

C c

or regiftered ; Scheme, the furrounding water fide ; Schifm, below the furrounding water fide ; Scholar, on the lowering or inclofing water place fide ; Sciatica, an ache at the fide ; Science, the knowledge of ens or exiftences ; Scion, the firft on the fide ; Scout or Scotch, or Scoff, leffening the inclofing water part ; Scold, at the inclofing water place fide ; Scope, the lower inclofing water part; Score, the fide border ; Scotia, the lower furrounding water or fea fide ; Scorn, leffening the inclofing water or border ; Scorpion, in the lower furrounding water or weftern part ; Scot, the lower furrounding water or fea fide ; Scour, the inclofing fpring water fide ; Scourge the fide fcouring ; Scout, below the fpring inclofing water fide ; Scrag, the fide or part of a rock ; Scramble, the furrounding water place or rock fide ; Scrap, Scrape, Scratch, Scrawl, Scream, Screw, Scribe, as refembling the divifion or border water fides ; Scroll, the inclofing water fide roll, as rollrich ; Scrub, the fpring part below the inclofing water part ; Scruple, the fpring below or at the fide of the inclofing water place; Scrutiny, the fpring on the inclofing water fide ; Scuffle, the lower inclofing fpring water place, as that betwixt the fpring or river and fea water at their confluence ; Scull, the upper covering place ; Scum, the furrounding fpring water fide ; Scurf and Scurvy, below the border water part ; Scuttle, the inclofing fpring water place fide ; Seyre emot court, of the earl and bifhop, the furrounding border or fhire meeting. The name Scotland feems to be of a Britifh or Saxon extraction of the fame fignification as the latin name Saxon ; at firft extended only to the inhabitants of the weftern and northern iflands and fea coafts of Scotland and the Irifh Seas; but afterwards like that of the Angles to Englifh in Southern Britain, it became the general name of all the inhabitants of North Britain But if the fouthern boundary of Scotland be fixed, as it ought to include all the invaders of Vertigern, at the Tyne or Hadrians wall, the proper limits of the Roman Briton ; the Scotæ comprehended the Scots, Saxons, Irifh Scots, and Caledonians, who from their fituation on the lower or weftern fea, were called Scots or the lower water fides or coafts, or the weftern water coafts from the fun's appearance of fetting in the weft ; and alfo of the Picts, the upper fide part, which from the fun's rifing in the eaft was deemed the upper part ; the Deucaledonians, the northern hills inclofing his fide or vales ; Vecturiones or Pchini, the upper fide highlands, and the Meatæ, Dalenieatæ and Ouadini, including the Gadeni, Selgovæ, Novantes and Damnii ; the former being of the Caledonian, Irifh, Saxon, or Britifh race, and the latter of the Cumberi, Dalematian, Gaulish, Welsh, English or Angli British origin. That the names Saxon and Scot, mean the same people both in Scotland and England also appears by the following Engli-Saxon law terms, viz. foc, foke, or fcot and lot or them; being the right of port toll and the liberty of landing and fcotal payable amongst the Saxons and Scots whose refidence were along the fea coafts, and the fair and market toll, lot or proportion payable to the Cumbri Gauls or English, who dwelt in vales on the boundary rivers, where fairs and markets were held under the fuperintendency of the heads of tribes or clans for the liberty of holding fairs and markets, and for taking public cognizance of all fales, bargains, and exchanges; Sac and Scavage being the like toll within manors or leffer diftrichs collected by the reve or fteward; Sakeber, below the inclofing water part, or bearing any thing beyond its proper limits; Soccage, a Saxon farm or the leffer diftricts; Sock-manni, the Saxon farmers; Scutage, below the inclofing water fide, or a tax in lieu of that fervice; efcheat, below its limits; Efcuage, the leffer inclofing water fervice, or the fervice of an efquire or leffer knight; and the various other Saxon and English, names of places here defined, which have been formed on the model of the generical or radical Sco or Scot, after the manner of other countries, without any derivatives from Sax or Saxon, except fuch as have been tranflated from the Latin. If we attend to what hiftorians have advanced touching the Scottifh origin, and refpecting the feveral denominations of the people, we fhall find this account of them in a great meafure confirmed, and that their origin, like the Saxons, has been derived from various countries and nations; as Tacitus, who fays that the Scots were Germans, probably the fea Germans, Sicambri or Gauls, the followers of Mercury, fince their ancient language appears to be the Celtic. Ammianus Marcellinus, who firft mentioned the Picts, fays that the Deucalidones and Vecturiones were comprehended under the denomination of Picts, but that the reft of Scotland were the Scoti and Attacoti, who together with the Franks and Saxons bordering on Gaul, were the invaders of fouthern Britain; Dion Caffius, who fays that the Scots confifted of the Caledonians beyond the great wall, and the Meate on this fide of it, to whom all the reft were related; Bede afferts that the Lowlanders were of the English origin, that the Scots came from Ireland, and the Picts from Scythia, meaning Scandinavia according to Rudbeck, Ufher, and Stillingfleet; that they fettled at the mouth of the Tyne, fpoke the Pictifh tongue, which by the by appears to be the Britifh, and that Britain received

Y   another

another different nation from Ireland, which explains what Bede meant by nations, since the Irish were of the first colonizing race of Britons; Pomponius, Aventinus, and others say that the Picts were Germans; the Scots historians widely differ as to their own origin, as favouring too much either the Gaulish or English or the old British origin, since they appear to derive their origin from every species of Britons. Humphry Llwyd places the Picts or Meatæ in Lothian. Camden and Edward Lloyd say that the Picts were the same as the Britons, and spoke the British tongue. John Picardus and some others derive the Picts from the Pictones of Gaul, in whose country the woad grew. Giraldus Cambrensis brings the Scots from the province of Ulster in Ireland, to the north of Scotland, in the fourth century, under Mured or Murry, but those seem to be Saxons. Tacitus describes the persons and ruddy complexions of the Calledonians to be similar to what Diodorus Siculus says as to the make and complexion of the Gauls. And Mr. Drummond of Hawthornden and others say that the Saxons were in possession of Scotland as far as Edinburgh till nine hundred and sixty, and that the Ottadini were called Scottadeni. And as the Scottish and Pictish customs, manners, language, names of persons and places, and their make and complexions agree with the Gaulish and British, those who bring the proper Picts from France, the proper Scots from the sea coasts of Britain and Ireland, the Calledonians from the proper hilly Gauls, and the Lowlanders from the Cumbri of this island, and the rest from an intermixture of those, seem to deserve the most credit; more especially if we regard the extensiveness of Gaul, and its numerous colonies, which passed the Rhine as well as into Britain, that the first migrations continued from the south northward, so long as the Gauls exceeded the Germans in strength and valour, and the country remained under the government of the Druids, and that the first colonies which came into Germany north about, must have been through Norway, or by sea, as it was otherwise shut up by the Hercynian forest. and if our ancient bards and writers may be credited, the British isles furnished Norway, Denmark, Germany on the Baltick coasts, and the Scandinavian empory with Britons and Cumbri or Gauls, in the colonizing age, and at the expulsion of the Druids by the Romans, who, during their residence in this island, retained the same vicinity betwixt the inhabitants of the several parts of Britain and those of the continent, as they had had on the continent in time past: the first migrations having been by sea, and the next in progressive streets from east to west thro' Thrace, Illyricum, Gaul, and Britain. It may not be improper

proper here once for all to obferve, that as the Greeks and Romans preffed the Celtic nations in Italy, Gaul, Spain, and Britain, they wheeled northward and returned fouthward into the Roman countries, and regained their former feats, as evidently appears by their change of feats, according to the ancient hiftorians and geographers. But what proves the Scots, as well as the Englifh to be a Celtic nation, beyond the poffibility of a contradiction, is their ancient voices and names of perfons and places being Celtic.

Seam, Axantos, or Samos in Greece, and an ifland over againft the Ofiffmi, or Britania Armorica, in the Britifh fea, the furrounding water or fea fide or ifland, both famous for their oracles and priefts called Lenæ, the water bank men or wardens. Seafdon, Staff. the lower or fide water fide town. Seaton, Devon, Kent, on the lower water or fea fide town. Seavanfhal, N-hum. the fide or lower inclofing fpring water part or haven high water place or hall. Hence Sea, the fide or lower water; Seal, the fide or fea water place, which marks the boundaries; Seam and Seon, a net or on the fea; Sear, the fea mark; Search, on the fea fide confines or action; Seafon, on the feas, which was in fine weather and the Saxons periodical time of going to fea; Seat, the fea fide or bank; Seaward, the fpring water at the fea fide, or the fea warden.

Secandunum or Seckington, War. a town on the lower inclofing water. Segelocum, Linc. the inclofing water place fide or ftation. Segedunum or Segehil', N-ham. the lower inclofing water place fide hill, town or ftation. Segantium or Setantium, Hants, N-Wales, Lanc. on the lower inclofing water or fea confines, ftation, fettlement or city. Segonax, Norf. on the lower inclofing water or fea confines, or Iceni chieftain. Segrave, the fea warden. Seymore, the lower furrounding more or fea. Seiont, on the fea fide. Hence Secant, the fea or lower inclofing water fide; Secede, below the inclofing water fide; Secude, below the inclofing fpring fide; Second, the fea or furrounding water fide; Secret, the lower or female inclofing water fide; Seel, below the inclofing water fide, or divided from the community; Secular, the fpring water place fide or country; Secunding, in the female inclofing water, or fecond part; Secure, on the border fide; Sedan, on the lower or feat part; Sedate, the fide or feat at; Sediment, at the bottom part; Sedition, on the lower or below the fide; Seduce, the lower or below the fpring part fide; Seed, the lefler or female parts or divifions; fee, Si or the fun's rays on the lower parts; Seek, the feeing action;

Seignior,

Seignior, on the inclofing water fide or country; Seizin, on the fea fide or in pofeffion.

SEL, Selaby, and Selby, Yorkf. Dur. the water place fide, part, abode, cell, grove, or low part. Selbrittenden, Kent, the water part fide low or woody vale, or the Britifh vale, cell, or grove. Selgovæ, or the inhabitants of Lideldale, Eufdale, Efkdale, Anandale, and Nidifdale, in Scotland, the dwellers on the lower inclofing fpring water confines. Selkirk, Scot. the lower border fhut, meeting, or church. Selfcomb and Selfey, Suff. the fea, falt, fide, or lower water place vale or ifland. Selwood, Som. Wilts. the fair, fide or lower water place wood. Hence Seldom, the fea water place furrounding border; Select, the fide water place fhut part; Self, the cell or fhut part; Sell, on the fide water place, which were the fairs and market places; Selvage, the fpring water place edge; Selon of land, the fea bank fide or lane, or a portion of land of the like dimenfions.

SEMPIL, Scot. the fide furrounding water place. Sempringham, Linc. the fea or lower furrounding water part confines, vale or habitation. Hence Semi, the border water fides or rim, or half the boundary water part; Semblance, feen on the water place, border, or fide; Sempiternal, the fide furrounding fpring part; Sempftrefs, the female at the feam part.

SENEN, an Ifland in the Britifh fea, the fea inclofed one; Sens, Senlis, and Senones, Gaul, the fide inclofing water place, and on the fide inclofing fpring water place. Senus, Ire. the fide inclofing fpring. Hence Senate, at the inclofing or inclofed fide; Send, the fide inclofing part; Senefchal, on the fide inclofing confines part; Senior, or an alderman, on the fide inclofing water man; Senfe, below the inclofing fide; Sent, to the inclofing fide; Sentence, below the inclofing fide; Sentry or Sentinel, on the inclofing fide part; Sentiment, the minds furrounding fight; Separate, at the fea water part; Septenary, on the fea fide part; Sepulchre, the fide water place covering.

SERKE, a Britifh ifland, the fea or lower water confines. Seria and Serphi, Greece, the lower or fide water part; Seraglio, below the water fide fhut; Seraph, the fon of the ftars or leffer lights; Serene, on the water fide, or the ftars in the other language; Serjeant, on the lower inclofing water fide man; Series, the fides to the water; Sermon, on the furrounding water or fea fide; Serene and Serum, the leffer fpring water; Serpent, the head along the ground or furface; Servant, the lower part of the houfe; Sefs and Seffion, on the

the lower water fide, where rates and taxes were affeffed and feffions kept for that purpofe.

SETANTII or Secontii, N-Wales, the lower fea fide, or the Saxon poffeffions. Setcia, Chef. and Spain, the fea fide or the lower poffeffions. Seton de la val, N-humb. the vale on the fea fide, or a fea town of the vale. Hence Set, the fea or lower fide. Seton, in the lower covering; Settle, the fea fide or lower place or the water place lower fide.

SEVERN, Gloc. &c. on the inclofing fide or feparating fpring water. Severcia, Wilts, the fide fpring water. Seven-fhale, N-hum. on the fide fpring lower place. Seven Burgenfes, burghs on the fide fprings, and Five Burgenfes, burrows on the fpring part, or perhaps in a remoter fenfe, burghs or diftricts of five and feven villages, fituated on the river fides, as Leicefter, Lincoln, Nottingham, Stanford, Derby, York, and Chefter, the Loegrian limits, where the Northumbrians and Scots of Fife made fome fettlements on the retreat of the Romans. Seven-oke, Kent, on the fide fpring water or feven oaks. Hence Seven, the fea on the fide fpring water, the haven, or refting or ftanding part, or a divifion in number; Seventy, feven fides, ties, or tens; Sever, the fide fpring water, which fevers; Severe, parting; Sewer, the lower water; Sex, the lower or female water confines, which makes the diftinction of fex, elfe all would have been neuter; Sexton, the lower water confines water fide one, it being the Saxon method to bury in water.

SHAFTSBURY or Septon, Dorfet, the lower fide part or burrow, or on the lower fide part. Shanon or Senus, Ire. the lower inclofing or fide high fpring. Shap, W-mor. the lower or inclofing part, formerly Hepe, the high part. Sharborn, Norf. on the lower or inclofing water border or neighbourhood. Sharpnor, in the Ifle of Wight, the lower water part on the border, or on the border edge. Shafton, the lower fide town. Shavington, Chef. the lower confines or inclofing fpring town. Hence Shaby, the lower habitation or covering; Shakle, a fhutting the lower place; Shade, at the fide; Shaft, the lower fide part; Shake, the fea or lower inclofing water, which is moft commonly in agitation; Shale, upon or inclofing the lower; Shall, the lower up; Shallow, a low water place; Shallop, the low water place thing; Sham and Shame, furrounding or covering an under part; Shank for Chank, the lower branch; Shape, the lower or fhade part; Shard, Share, Shark, and Sharp, the edge fide or lower water part; Shatter, and Chatter, at the lower part, or ftriking at the lower part; Shaw, the lower fprings; Shave, the cutting thereof.

SHEAFIELD,

SHEAFIELD, Yorkf. the fide or lower water dwelling place or village. Sheales, N-hum. the lower water place fide. Sheerneſs, Kent, the fide or lower water promontory, or iſland. Shelford, Nott. Bedf. the fide or lower ford. Shelſey, Heref. the lower fide water place. Shene, Surry, on the incloſing or fide water. Shepey, Kent, the lower or iſland part. Sherry-hutton, Yorkf. on the fide or incloſing water high fpring part or town. Hence She, the fide, lower, leſſer diviſion or female he, hi, or i: Sheaf, an incloſed or fhut part; Shear, upon the fheep; Sheath, the incloſing part; Shed, fheen or on the fide; Sheep, the lower part; Sheer, the leſſer water; Sheet, the lower covering; Shelf, a low water place part; Shell, on the lower part or water place; Shelter, on the water place fide, or a low incloſure; Shelving, the lower water place edge; Sherif, the fhire reve or warden.

SHIN, Chin, or Kin, Scot. the fide incloſing or foremoſt water part. Shingle hall, Herts. below the incloſing or Ingles high water place or hall. Shipton, War. Salop, the fide part town. Shirburn, or Scireburn, Dorſet, Oxon. Yorkf. on the fide, border, or incloſing fpring water brooke, neighbour-hood, or fhire. Shirwood, Nott. the incloſing fpring water fide or wood, or an incloſed or covert wood. Shirley, the fhire or border fpring water place. Hence Shield, the fide or water place fide covering. Shift, the lower or female part covering; Shilling, on lower incloſing or Engliſh water place fide; Shin, on the fide; Shine, the fun in or on the fide; Ship, a fea thing; Shire, the border incloſing water; Shirt, the lower covering; Shiver, a fevered part.

SHOAD, Corn. the fhew at the water fide. Shobery, Sceobirig, Eff. the furrounding water part. Sholache, Chef. the lower furrounding water place confines or fhut. Shorcham, Suff. a vale or dwelling place on the furrounding water, fea or fea fhore. Shom, Bedf. on the furrounding water or border fide. Shotwick, Chef. the furrounding water fide ſtreet or fortification. Hence Shoal, low or fhallow water place; Shock and Shog, the furrounding water or fea confines; Shoe, the lower covering; Shot, at the furrounding water fide; Shop, the furrounding water or fea fide part; Shore, on the furrounding water; Shorn, on the furrounding water; Short, at the furrounding water; Shove, the furrounding water fpring; Show, the furrounding fpring; Shower, the furrounding fpring water.

SHRAWERDAN for Chrawerdan, Salop, at the border fpring fide. Shrewfbury or Shrowfbery, Salop, Som. the furrounding or border fpring water part or burrow. Slughbury, War.

War. the fide fpring water burrow. Shurland, Kent, the fide fpring water bank fide or land. Hence Shred and Shrew, dividing or parting; Shrimp, the leffer or lower water border part; Shrine, in the lower water; Shroud, the lower furrounding water covering; Shrove tide, the fpring tide fide; Shrub, the leffer fpring part; Shrug and Shudder, the water part fide; Shun, below the fpring; Shut, the lower fpring fide; Shy, the low.

Siadæ Iflands, in the Britifh fea, the fide or lower water parts. Siambis, another ifland in the fame fea, the furrounding fide water lower or ifland part. Sibertofte, N-ham. the fide water part lower border coaft. Sicily, the fide water place ifland. Sidenham, Surry, a vale on the fide or lower part, or the wood fide vale. Sidmanton, Hants, the fide or fea part town. Sidmouth, Devon, the fide or fea mouth or gate. Sidnacefter or Stow, Linc. a fortification, ftation, or city on the fea fide. Sidney, Kent, on the fea fide. Sigdeles, Sillinæ, Silures, Hefperides, or Scilly iflands, in the fea or fide water place, or the lower high water part fide iflands. Silbury, Wilts, the fide water place fpring part, dwelling place or burrow. Silt, Linc. the fea or lower water place fide. Silures, Wales, below the fide fpring water place or Severn. Silurnum or Silchefter, Hants, on the fide fpring water place city. Sinnodunum, Berks, Yorkf. a town or hill on the fide water or fpring water part. Sifingherft, Kent, below the inclofing water or fea fide foreft. Sitomagus or Thetford, Suff. the lower border furrounding water way, ftreet or city. Sittingburne, Kent, on the lower inclofing fpring water fide part, neighbourhood, or ftation. Siward, the fide or fea warden. Hence Siccity, the lower inclofing water or fea fide; Sick, low; Side, the lower or fide divifion or the rib or rivulet part; Siege, below the border or edge; Sift, the chaff off the growth; Sigh, a low hie; Sight, the lower fire or light part; Sign, on the fide water or the meres; Signify, the fign part; Silence, the found on the water place fide; Silk, the upper fide covering; Sill, the lower place; Silly, the leffer light; Silver, the lower or leffer light fpring, perhaps the moon's, the origin of money and mine, as gold ore is the fun's property; Similar, the border fpring water fide; Simper and Simple, without art or part; Sin, in or on the fide water fide or infide; Since, below the infide or part; Sincere, a warm infide heart, or dear; Sinew, an infide fpring; Sin an infide found; Single, an ifland or inclofing water place; Sruttle, the lower inclofing fide water part; Sink, a low that; Site, a fide inclofed; Sip, a fmall part; Sir and Sire, a title of a knight;

fixth,

Sifter, the lower or female fide part; Sit, the lower fide; Situate, at the lower water or fea fide; Sixty, the fea or water confines fide or tye; Seize, covering the lower.

SKEFFINGTON, Norf. a town on the lower inclofing water part. Skelton, Norf. a town on the inclofing water place. Skellingthorpe, Linc. the inclofing water place fide gate. Skengrave, the inclofing water fide warden. Skeweth, Wales, the lower inclofing fpring water fide. Skiddaw, Cumb. the fpring or fky water part or fide, or fpringing up to the fky. Skinfrith, Wales, on the lower inclofing water fide way. Skipfay, Yorkf. the lower inclofing water or fea part. Skipton, Yorkf. the lower inclofing water part town. Skipwith, Leic. the lower inclofing fpring water fide. Skirlaw, the border fpring water place fide or law. Skrckingham, Linc. a vale, home, or dwelling place on the lower water border. Hence Skein, the lefter ties, binds, or fhuts; Skate, at the lower water or fea; Skeleton, the lower fhell covering; Sketch, the lower inclofing water fide; Skiff and Ship, the lefter or fide water fhut or veffel; Skill, on the inclofing water place; Skim, the lower water border or covering; Skin, on or covering the lower; Skirmifh, below the furrounding border fide; Skirt, the lower border fide; Skreen and Skulk, on the border fide or inclofing water place; Sky, covering the lower parts.

SLANE, Ire. the fide water bank, or on the lower water place. Slaugham or Slaughden, Suff. the fide or lower water place vale. Seaford, Linc. the lower or fide water place, ford or way. Slego and Sligah, Ire. the lower inclofing water place. Slew Gallen, Ire. the Englifh or Gaulifh inclofing fpring water place or mountain. Slingfby, the inclofing fpring water place fide, part, or habitation. Hence Slab and Slate, a broad or flat fide part; Slack, the leaft water place action, or a low lake; Slander, leffening the water bank part; Slant, the fide water bank; Slap, below the hand, or an high part; Slave, a low living place; Slay, lay low; Sled, at the water place fide; Sleck, the lower water place fhut; Sleep, the lower place or water place part; Sleet, the lower or icey water parts; Sleeve, below the hand part; Sleight, a low water place fide; Slender, the low water part bank; Slice, leffening or dividing the fide water place; Slide and Slip, along the divifion water place; Slim, the lefter border water place; Sline, the furrounding water place fide; Sling, the lefter inclofing water place; Slit, the lefter fide water place; Sliver, the fide fpring water place; Sioats, Slobber, Sloe, Sloop, Slop, Siope, Sloth, Slouch, Sloven, Slough, Slow, Sludge, Slug, Sluice,

Slumber,

Slumber and Slut, the furrounding water place fide or the lower or leffer furrounding water part; Sly, the fide or below the water place.

Smalbridge, Suff. the leffer middle or divifion water place bridge. Smerwick, Ire. the lower middle or furrounding water or fea confines ftreet or fortification. Smere, Yorkf. the lower or leffer furrounding fpring. Smithy Rent or Bloom Smithy Rent, the fea furrounding fide or coaft fmelting rent, or the houfe or Lord's part. Smyrna, in Ionia, on the lower fea or furrounding water fide. Hence Smock, the tide furrounding confines, meeting or killing; Small, on the fide furrounding water place; Smart, at the furrounding water fide; Smear, below er at the fide of the middle water; Smell and Smelt, on the furrounding water place fide, the ufual place of ore melting; Smerk and Smile, the furrounding water place fide or fight; Smite, the furrounding river or water at the fea; Smith, the fea furrounding tide, the ufual places of fmelting; Smock, a covering about a female; Smoke, the leffer furrounding water or fluid; Smooth, the leffer furrounding rdes; Smuggle, below the furrounding fpring water place; Smut, at the the furrounding fpring fide.

Snailwell, Suff. the fide inclofing fpring water or dwelling place. Snath, Yorkf. the fide or lower inclofing part. Snaudonia or Snowdon, N-Wales, on the lower inclofing water fide, hills or fnowy foreft. Snaite, Nott. the lower inclofing fide water. Snothill, Heref. the leffer or lower furrounding water fide hill. Hence Smack, the fide inclofing water confines or edge; Snaffle, an inclofing part upon; Snag, from the inclofed part; Snail, a low or flow inclofed place; Snake, the fide inclofing water confines; Snap, a fide inclofing part; Snaphick, Snare, the fide inclofing water or fhutting on the lower; Snatch, at inclofing or catching; Snear, on the fide of the inclofing water; Snip, the fide inclofing part or thing; Snivel, the leffer inclofing fpring water place; Snore, Snort, Snot, Snuff and Snuffle, the lower inclofing water fide, or a nofe at; Snow, the leffer inclofed or thin fpring water; Snub, below the inclofing fpring part; Snug, the leffer or lower inclofing fpring water confines.

Soar or Soer, Leic. the furrounding water or forfe water. Socburn, Dur. the furrounding water neigh low or dwelling part. Sodbury, Glouc. the furrounding water fide fpring part or barrow. See allo the water or furrounding water part. Soham, Camb. the fea or furrounding water vale. Soke Liberty, Edf. the fea or furrounding water confines or Saxon liberty. Soland Geefe, the fea lands nue geefe. Solent, the furrounding

furrounding water fide men. Solenfian, on the furrounding water bank. Solente, Hants, the fea or furrounding water bank fide. Solonia or Colonia, the furrounding water banks or lanes. Solway, Scot. the fea or furrounding water place way. Solyhill, War. the furrounding water place hill. Somerley, Suff. the fea furrounding water place. Somerfet and Somerton, the lower furrounding fea fide, feat, fituation or town. Somerfham, Hunt. the mere furrounding water fide vale. Somerville, the furrounding mere dwelling place or village. Hence Some and Something, the furrounding water fide ; Son, in the furrounding water part ; Soon, on the furrounding water part ; Soot and Sooth, the furrounding water fide ; Sop, Soph, Sophi and Sophift, the furrounding water part fide ; Soc and Sithefoca, the fea fide or Saxon limits or borders, with the liberties of a court, like the leet and baron courts of the Gauls in the inland countries ; Sockmanni, the fuitors thereof.

SORBODUNUM, Corbodunum or Salisbury, Wilts. the furrounding water or border refiding town. Soure and Soureby, Nott. Yorkf. the furrounding fpring water or a port or habitation thereon. South, the furrounding fpring fide. Southam, Southampton, Southwark and Southery, Devon, War. Hants, Surry, the furrounding water fide vale, vale town, ftreet fortification or country. Sow, Staff. the furrounding water fpring. Hence Sorcerer, the furrounding water border man or keeper ; Sordid, at the furrounding water fide or poffeffions ; Sore, on the furrounding part ; Sort, the furrounding water fide port or poffeffion ; Sofs and Soufe, below the furrounding water fide ; Sot, at the furrounding water part ; Sovereign, on or over the furrounding water confines ; Soul, the furrounding fpring of light ; Sound, the furrounding fpring in the parts ; Soup, the furrounding fpring water part ; Sour, the furrounding fpring water : South and Sow, the furrounding fpring fide and water animal ; Soil, the furrounding place.

SPALDING, Linc. the lower inclofing water place fide. Spaldwic, Camb. the lower water place fide ftreet fhut or fortification. Spaws, the lower fpring parts. Spelhurft, man and well, Kent, Norf. Glos. the lower or fide water place wood, foreft, part, fpring, or dwelling place. Spene, Spinæ, Spincham or Newbery, Berks. on the fide water part, vale or habitation. Spey, Scot. the lower or fide part or water part. Speney Caftle, Scot. a caftle on the Spey. Spurhead or Ravenfpur, Yorkf. the fide or lower fpring port head or the river or fpring lower part. Hence Space, the lower fide part ; Spade, below the foot part ; Span, the fide of the higher or upper paw or hand ; Spaniel, on the water part race ; Spanifh, the

lower

lower part fide; Spar, the under or lower bar; Spare, a low or bare part; Sparkle, the lower or leſſer fire part light; Sparrow, the thatch animal; Spawl, the lower part ſpringing up; Spawn, the leſſer ſpringing diviſions; Spay, leſſening the part; Speak and Speech, the ſide or ſound part action; Spear, on the lower or ſide part; Special, on the lower ſide part; Speck and Species, the leſſer incloſing water parts or a combination or union of parts; Speculum, the ſide or light ſpring part; Speed, the lower part at; Spell, on the lower water place part or wall; Spend, leſſening the inſide or incloſed part; Sperm, the lower furrounding water; Spew, ſpringing up from below; Sphere, the earth or on the lower ſide; Sphinx, below the incloſing water part or the Nile ſide, which were often attacked by its floods, and nothing but the interpretation of its riddles, or making good its banks and opening it could avoid its invaſion, which Oedipus performed by opening its way through a rock; Spice, raiſing a low part or thing; Spike, the leſſer pike; Spill, below the water place part; Spin, the lower part on or incloſing; Spinſter, one who has not ſpinned; Spiny, the leſſer pin; Spire, the ſmaller part upwards; Spirit, the leſſer or ſmaller fire or light part or diviſions; Spiſſitude, at the lower ſide part; Spit, the lower or ſide part at; Splaſh, below the water place ſide; Splayfoot, the ſide place foot; Spleen, the ſide water place within; Splendour, on the water ſide place; Splice, below the ſide water place; Split, the ſide water place diviſion; Splutter, the leſſer place utter or ſpring part; Spoil, below the furrounding water part place or race; Spoke, the leſſer furrounding water part confines or ſide; Spondee, the lower ſound part in two or diviſion; Sponge, the leſſer furrounding water part ſhuts, ſhells or pores; Spoon, a water veſſel in the hand; Sport, the ſide furrounding water part, uſually wilds, woods and foreſts; Spot, the leſſer furrounding water part ſide, part or poſſeſſion or leſſer circle part diviſion; Spouſe, the ſide part of us; Spout, the ſide water part out; Sprain, the ſide part from in; Sprat, the leſſer at the water part; Sprawl, the lower part from high; Spray, on the water part ſide or poſſeſſion; Spread, the water part ſide or country leſſer parts or feeds; Sprig, the leſſer water part confines, or a ſmall branch of a river or tree; Spright, the water ſide ſpirit; Spring, the incloſing water lower part, the fire at the lower parts, or the lower parts or ſurface; Sprout, the lower part out; Spume, furrounding the water part ſide; Spur, a low or ſide part ſpring; Spurious, the leſs pure, ſpring or off-ſpring part; Spurn, on the ſpur; Spurt and Sputter, at the ſpur; Spy, the feeing thing; Squab, the leſs or low ſpring part, as a chick, or a low fall; Squall, the lower

water

water place or fea fpring; Square, the inclofing fpring fides
Squath, the lower inclofing water fide; Squat, at the lower inclofing water confines; Squeak, the leffer fpring water confines
or action; Squeeze, below the inclofing water fide; Squib,
fpringing up from below; Squire, the leffer on the inclofing
water; knight or kingt being the chief, firft, king or doge thereof; Squires and Squirt, fpringing up from below.

STAFFORD, the lower, Iceni or Ikenild coaft way or high
way or ford. Stamford or Hanford, Devon, N-ham. Linc. the
furrounding fide or coaft way or ford. Standifh, Lanc. the
lower inclofing fide coaft. Standon, Herts, a town on the
lower or Iceni coaft or confines. Stanhope, on the fea or
lower high water part coaft. Stanmore, Yorkf. Midd. on the
lower coaft or fide moor. Stanes, Midd. on the lower fide
coaft. Stannaries, Corn. on the lower coaft country, Stannum
and Stone. Stanfted, Eff. on the lower coaft or Iceni ftation.
Stanthorp, Dur. on the lower coaft water gate part. Stanton
Drew, Som. a town on the lower coaft fpring water part.
Stanwell, Midd. on the lower coaft fpring water place. Stan-
wick, Cumb. Yorkf. on the lower coaft ftreet, fhut, fortification or city. Staple, the fea coaft ftanding or ftation place.
Stephelm ifland, in the Severn, the lower fide part ifland.
Sterburrow, Surry, the lower or fide fpring water part or burrow. Sterling, Scot. the lower inclofing water part place.
Stert, Devon, the lower water fide part or point. Steven-
haugh, Dev. the lower inclofing fpring water part or ftation.
Steward, the under, lower or leffer fide, poffeffions or houfe
warden. Stilton, Hunt. the lower fide water place town.
Stiperftone, Salop, the lower fide water part ftones. Stoke
and Stow, Scot. Nott. the lower border water fide, key or
ftation. Stokeley, the lower border ftation place. Stoke Gomer and Stoke Carey, Som. the fea and fpring water lower
border ftation. Stoker, Oxon. Scot. on the lower water border. Stonehenge, Wilts, the high inclofing water or kingfth
border, ftile or ftones as they were the meres and landmarks
between the ... Germans, Legarian Saxon, Iceni or Faft
Angl ... and the ... ry, litering the feveral conflicts of the
country, where to ... ly the meetings were annually held of the
druids, in fuch places, and afterwards of the heads of the Clans,
... ... buildings. Britons were muftered by the Saxons
... ... in council, for adjufting matters, and the border
... ... good Terminus, according to the
... ... of Sir Inigo Jone. Stondley,
Hunt ... ... the lower border place, palace or ftation. Stony-
... ... the lower border ftreet or ftation. Stort,
Herts.

Herts, the lower or Iceni border water part. Stortford, the
Stortford way or ftation. Stour, Dorfet, Wilts. Kent. Eff.
Suff. Worc. the lower border or ftation fpring water part.
Stour-mere-minfter-mouth-ton, Suff. Dorfet, Kent, the border
fpring water fide lake, monaftery, mouth or gate and town.
Stowey, Som. the lower border fpring. Straban, Ire. on the
lower river water part. Strangford or Strangway, Ire. Dur.
the inclofing water part or fea fide way, ftrand or fortification.
Strat, Stratth or Stret-clwyd-fieur-ford-bolgy-divern ern-navern-
ton-well. Scot. Bucks, Salop, Staff. War. Linc. Bedf. Corn.
the water part way or ftreet, on the inclofing fpring water
place fide, the furrounding water place confines, divifion fpring
inclofing water or fpring town, and fpring water, well or
dwelling place or the principal ways, roads or ftreets, fuppofed
to be Roman, but more probably from their names and thofe
of their inhabitants, which the Romans latinized, the ftreets on
the different borders of countries. Strelley or Stradley, Wilts,
Nott. on the lower fide water place or ftreet. Strigil or
Chepftow, S-Wales, the lower inclofing water part fide place
or key part. Stubham, Yorkf. the lower fide fpring part vale
or houfe. Stuccia or Iftwyth, S-Wales, the lower fide fpring
water. Studley, War. Wilts, a place at the lower fide fpring
or a ftream place. Sturbridge, Camb. Worc. the lower fide
fpring water bridge. Stutfall Caftle or Lemanis, Kent, a caftle
at the lower fide water place or the water place on the furrounding water part or haven or the German part. Hence
Stable, ftanding place; Stack, Stag, Stage and Stagnate, covering the lower part; Staid, Standing, Stain, in the lower
part; Stair, the lower part high or up; Stake, the lower inclofing water; Stale, a ftanding water place; Stalk, the lower
or fide water place confines; Stamp, the foot about the lower
part; Stanch, ftanding on the inclofing water confines; Stand,
or an end; Stannary, on the lower coaft country; Staple, the
for or lower coaft ftanding or ftation place; Star, the ftanding
fire, ftuck, ftanding on the inclofing water confines; Star
Chamber, the ftate chamber; Stave, upon a ftand; Stark, ftanding on the inclofing water; Stay, ftanding on the inclofing
water place; Start, ftanding on the water fide; Starve, fide on
a ftand; State, the ftanding fide part or pofition; Statute,
the ftanding ridge; Stay, the ftands Steall, the ftanding or
ftation part or water part; Steel, while or at firm; Steep,
low the water place fide; Steer, ftream-ing; Stem,
Steep, at the lower fide or ; Stew, ftanding
place; Steps, the ; Stern,
Stib; Stem, the lower fide water; Stich, on

clofing water fide ; Step, from ftanding ; Sterii, a ftanding fpring or flowing water place or race ; Sterling, the lower country inclofing water, English or ifland place, and thence Sterling reckoning of money fignifies Englifh currency ; Stern, on the waterfidefland ; Stew, the ftanding fpring ; Steward, the lower or under houfe warden ; Stick and ftich, below the inclofing water confines ; Stiff, the fide flowing part ; Stile, below or under the fide water place ; Stigma, below the furrounding water fide ; Stile, below the water place fide ; Still, on the water place fide ; Stimulate, at the furrounding water place fide ; Sting below the inclofing water place fide ; Stink, the lower inclofing water confines or back fide ; Stint, the lower, or leffer inclofing fide ; Stipend, under the head of the houfe or poffeffions fhare or divifion ; Stir, to the lower fide ; Stive, the lower or under the fide or houfe ; Stole, the fide covering upon ; Stomach, the fide furrounding confines ; Stone, on the lower border ; Stool, the fitting place ; Stop, the ftanding or lower border part ; Store, on the lower border or covering ; Storm, the lower border furrounding water ; Story, on the lower border ; Stove, the water covering part ; Stout, the fide fpring water ftanding part ; Stow, the border or inclofing water fide or ftation ; Straight and Strict, the water part confines ; Strain, the leffer inclofing water part. Strand and Strange, on the water part fide bank or river, as on the Thames and the fea fhore ; Strap, the leffer fide fpring water part ; Stratum, Straw and Strow, below or under the fpringing part ; Stray, below the fide water ; Streak, the fide water part confines ; Stream, the fide furrounding water part ; Street, below the water part fide ; Strength, the lower inclofing water part confines ; Strefs, the lower water part fide ; Stretch, the lower water part inclofure ; Strew, below the fpring water fide ; Stride, the lower water part fide or divifion ; Strife, the lower fide water part ; Strike and Stroke, the lower water part fide confines ; String, the lower inclofing water part or boundary ; Strip, Stripe, Strive and Struggle, the lower fpring water part up ; Strol, below the furrounding water part place below, the lower furrounding water part confines ; Stronger, the lower furrounding fpring part or thing ; Stud, the lower water part at the fpring or well ; Stub, the lower or fpring part ; Stubborn, the leffer fide fpring part or border ; Study, below the fide fpring part ; Stud, the lower fpring ; Struddle, below the more fhore roling fpd water fide ; Stump, the lower or fide ground place ftem, on the lower fide ; Sturdy, below ; Sught, below the water .... fpring ;

fide fpring water part at ; Sturdy, the fpring water part at the fide ; Sty, the lefter or lower houfe; Stygian, on the lower fide confines ; Style, below the water place fide or on the country.

SUARDONES, Germany, on the fpring water fide. Subdinum, Gaul, below the lower fpring part. Sabus, a river of Fefs in Africa, the lower or fide fpring part. Sue, Ire. the fide or lower fpring water confines. Sudbroke, S-Wales, the lower fpring fide brooke. Sudbury, Suff. the lower fpring fide burrow. Sudcote Steel, Yorkf. the lower fpring fide at the furrounding water confines ftation place. Sudley, Gloc. the lower fpring fide place. Suer and Suerby, Eff. the fide fpring water part. Suefliones, Belgic Gaul, on the lower fpring fide. Suffolk, the lower furrounding fpring water place confines, or the fouthern folk; the northern being the Northumbrians or northern Chambri. Suire, Sewer, or Swiliv, Ire. N-Wales, the fide fpring or water place. Sul by, Eff. the fide fpring water place, abode or dwelling. Sulloniacæ, Herts, or rather Culloniacæ, the fide or inclofing fpring water confines, banks, confines, lanes, ftreets, vales, or Watling ftreet, which, notwithftanding the Roman difule, feems to be the feats of the ancient Gauls, of the Comatus, or dwellers in vales on the river fides, and the firft gainers and dividers of countries by rivers into comots or cantons, whofe firft migrations feem to have been from Spain to the Vistula along the fea fide, gradually extending their boundaries. They proceeded weftward into Britain and Ireland through Gaul, where fome of thofe Cumbri remain to this day, intermixed with others of the fame origin from Denmark, Germany, Gaul, Dalmatia, Northumberland, Cumberland and other places. Sumerford or Cumerford, Wilts, the fide fpring water, way, or ford, or the Cameri or Coombes way of migration in comots on the river fides. Sundaland, Ire. the lower inclofing fpring water part bank fide, land, part, or country. Sunici of Belgic Gaul, on the lower inclofing fpring water. Suning or Suningwell, Berks, on the inclofing fpring water fide or dwelling place. Surlecy, N-Wales, on the fpring water lower fide. Surry, on or below the fpring water. N-Wales, the lower fpring. lower water confines, or fide. lower water fide confines. Roman or Sulfonic. Suerta, Scot. the lower fpring water bank the land or country, or the middle or divifion. Sutton and Sutterton, Lanc. Nott. and feveral other parts, a town on the lower

er

er fpring or river fide. Suevi, the fpring fide dwellings. Hence Suafive, the lower fpring fide dwelling; Subdivide, the fide or lower fpring part divifion or leſſer parts; Subdue, from the fide or lower divifion fpring part; Subjoin, the fide or lower fpring part go in; Sublime, the lower border fpring water place, palace, or leet; Submerge, the fide fpring furrounding the water confines; Submit, below or under the furrounding fpring fide or country; Subordinate, below or under the fide fpring part circle; Subfide, the lower fpring part fide; Subfidy, the fpring fide or confines part; Subſtance, under the fpring part; Succeed or Succefs, at the lower fpring confines; Succulent, the lower fpring confines bank fide; Such and Suck, the fide fpring water confines; Sudden, below the lower fpring water fide; Sue, the fide fpring; Suffer, the fide fpring water part or ordeal; Suffice, the fide fpring water part flowing; Suffocate, the furrounding fpring part fhut; Suffrage, on the inclofing fpring water part or country; Sugar, fpring inclofing water; Suit, at the fide fpring; Sullen on the fide fpring bank, Sully, on the fpring fide place; Sultry, the part on the fpring or fouth fide; Sum, the furrounding fpring fide; Summer, the furrounding fpring, or chief fire or hot divifion part; Summon, on the furrounding fpring part; Sumptuous, on the furrounding fpring part fide; Sun, the fpring one, or on the furrounding fpring one; Sup and Supper, the fide fpring water part; Superfine, the fide fpring water part confines; Superfluous, the fide fpring water part flowing the fides; Supple, the fide fpring place; Supplement, the inclofing fpring water place fide part; Supply, the fide fpring part water place; Support, the fide fpring furrounding part; Suppofe, below the fide fpring furrounding part; Supprefs, below the fide fpring water part; Supreme, over the fide fpring water furrounding part or country; Sure, on the fide fpring; Surety, on the fide fpring fide, tye, poffeffion, or property; Surge, the fide fpring water confines; Surely, on the fide fpring water place; Surmount, the lower fpring water, on the furrounding water part; Surname, the fide fpring water or country name; Surpafs, the part below the fpring water fide, or the river paffage; Surrender, on the fide fpring water part; Surround, the fide fpring water round; Survey, the fide fpring water view; Sufpend and Sufpence, the fide fpring below the end part or fide; Suftain, the part below the fpring part; Sutler, on the fide fpring water part.

SWAFFHAM, Norf. the fide fpring water part vale or dale. Swafton, Linc. the fide fpring water part town. Swaine Mote, on the fide fpring meeting. Swale, Kent, York. the fide fpring spring water place. Swallow, Surry, below the side spring water place. Swanescombe, Kent, on the side spring water side or lower vale. Swanpole, Linc. on the side spring water pole or surrounded water place; Swaneswellpole, War. the inclosing spring water side, spring water place, or well pole. Swansey or Abertau, S-Wales, the side spring on the sea, or the spring coast water part or harbour. Swift, Leic. the lesser side flowing spring. Swinbern, N-ham. the neighbourhood on the inclosing spring. Swinton, Yorks. a town on the side spring. Swonon, Kent, the side spring water surrounding parts. Swoninzil, Corn. a dwelling place on the side spring. Hence Swab, the side spring water part; Swagger, the side spring water confines; Swain, on the spring side or a youth; Swallow, Swamp, and Swan, on the side spring of water or animal; Swao, the side spring water part; Sward, the low springs on the side or the surface; Swarm, the small surrounding springs of animals; Swarthy, the southern country side; Swath, below the side spring; Sway, the side spring water; Swear, on the side spring; Sweat, at the side spring. Sween, the side spring part; Sweet, at the side spring; Swell, the side spring on the water place or sea; Sword, the side spring water side; Swift, Swig, and Swill, the side spring on the water place; Swirl, the side or the lower surrounding spring; Swine, below the inclosing spring; Switch, the side spring at the water confines; Swivel, the side spring well; Sword, the surrounding or edge spring water part; Sworn, on the surrounding spring part; Syllable, on the coil or incl... or side water place or ... Symbol, on the side surrounding spring part; Symmetry, the side surrounding ... ci-vilions or measure; Sympathy, ... side surrounding spring part; Synod, together ... circle; Syntax, the side surrounding spring tide, proper laws, positions, ties, duties, bonds, &c.

TACHMELIN, Staff. on the middle or inclosing water place confines or coast. Tackley, Oxon. the side covering water place. Tadbury, Tadcaster, and Tadmarton, Oxon. Yorks. the side covering spring part, burrow or dwelling place, city or more town. Taf, Taif, Tave, Tau, Taw, Tay, Tees, Teifi, Teivi, Teave, Tees, Teie, ... and Tobias, the names of boundary rivers of Corn. Devon, Wales, Yorks. &c. and other part, signifying the side covering division or border spring, or flowing water part ... ... Tadburrow, Norf. the lower side covering, confines, or liberi burrow. Tailponte, N-ham. the side covering or confines of the surrounding

A a   rounding

rounding fpring water parts, diſtricts, or wapontache. Tai-zali or Taxali, Scot. the water place fide covering or confines. Talbot, the part or coaſt above the furrounding water part fide, or habitation. Talcharn, S-Wales, the part above the high incloſing water place. Talgarth, N-Wales, the part or coaſt above the incloſing water fide. Tal-lyn, N-Wales, the part or coaſt above the lake. Tame, Tamer, Tamera, Taum, Teme, Teind, Tamefis, Thames, Corn. Oxon. Worc. Staff. Shrop. Berks, Midd. Eſſ. &c. the fide furrounding, incloſing, or co-vering fpring part. Tamerworth, Corn. by the furrounding fpring water part fide. Tamworth, Staff. the furrounding fide fpring water or dwelling part. Tanatos, Tanet, or Tha-net, Kent, N-Wales, on fide coaſt or covering water part. Tanfield, Yorkf. the fide or lower covering dwelling place. Tania or Dania, on the water fide coaſt, or lower part. Ta-niſtry, the Daniſh or the water part fide tenure. Tantallon, Scot. below the covering or water bank coaſt. Taodunum, Scot. on the covering fide fpring or Tay town. Tarbar, Scot. at the covering or boundary water part. Tarberth, Scot. at the covering or fide water parts. Tarf, the covering water part. Tarvedinum, Scot. the covering fpring water part. Tar-rent, Dorfet, the covering water on the fide. Tafe, the lower covering growth, or the limits or part demanded. Tatterſhal, Linc. the lower water part covering. Taviſtoke, Devon, co-vering fpring water confines or ſtation. Taunton or Thone-ton, Som. a town on the covering fpring or Thone. Taurica Cherfonefes, the upper covering fpring water on the lower bor-der. Tauton, Devon, on the covering or border fpring, or the fpring fide town. Taximagulus, one of the chieftains of Kent, the furrounding covering or fide furrounding water part, or the fea coaſt chieftain. Hence Table, the covering place; Tacit, covering the fides or found parts; Tack, co-vering water confines, which tacks the parts together; Tad-pole, covering at the pool or water place fide; Tag, the cover-ing ſhut; Tail, the covering place: Tailor, the covering place man; Take, covering or incloſing the water confines; Tale, on the covering or confines place; Talent, on the covering water place fide or bank; Talifman, the covering water place fide man; Tall, at the covering or fide water place; Tallage, on the fide or covering water place edge; Tallow, covering on an animal infide; Tally, the covering water place, whofe fides tally; Tallyman, the covering water place fide man, where fales were made; Talon, upon the covering; Tame, the co-vering or furrounding water fide; Tamper, covering the part; Tan, the fun heat or high covering; Tang, on the covering

water

water confines; Tangent, on the high covering fide; Tangible, the covering place; Tank, a covering on the water; Tanner, on the covering water or fkin man; Tantalize, on the lower covering water place fide; Tap, the covering part; Taper, the covering water part; Tar, the covering water; Tardy, the covering water fide; Target, the covering water fide or fhut; Tarras, a covering on the furface; Tarry, on the covering water fide; Tart, Tartane, and Tartar, the covering water fide; Tafk, the lower or under water covering action; Taffel, on the lower covering; Tafte, the infide covering; Tatter, the covering at the part or in divifion; Tavern, on the spring water covering or border; Taunt, on the covering spring water fide; Taw, an animal covering; Tax, from the covering or furface of the ground.

TEGANIA or Tegangle, N-Wales, on the water covering, or the Englifh water confines; it may be here obferved, that te differs in fenfe from ta only as it is the feminine, paffive, and fluid gender of ta the mafculine, active, and fubftantive. Teifidale, Scot. the Teifi vale. Teign, Teine, Tern, Tering, Thet, Thone, Thyrn, Tayne, Tina, and Tine, Som. Salop, Staff. Suff. Norf. N-hum. Scot. the fide inclofing or fpring water. Temsford, Bedf. the fide furrounding water way or ford. Tenbigh, S-Wales, the part or habitation on the fide water confines or coaft. Tenderden, Kent, on the water part fide vale or foreft. Teringham, Bucks, the inclofing water fide vale. Telbury, Gloc. a burrow at the fide water. Tetnal, Staff. a place on the fide water part. Tethys, the wife of Neptune, the lower fea or female fide, and thence Teat or Bubby, and the Welfh Teth. Teu or Tew, Oxon. the fpring fide. Tewkfbury, Gloc. the fide fpring water fide burrow or dwelling place. Teutobochus, Teutomarus, and Teutones of Germany, the fide fpring water border chieftain, mere man, and inhabitants, who dwelt along the German fide of the Rhine, and were in the fame alliance with the ancient fea Germans, as the Gauls or Cymberi were with the Britons, that is, brethren of the fame country or nation, and coufins of the Cymberi, as dwelling on the fame border water confines, according to the meaning of the word couzens, as well as their defcent from the old Germans, who were brothers of the ancient Britons, as being in all their fea-coafting migrations of the fame countries, and the Teutons having accompanied the Gauls or Cymberi in all their wars with the Romans, and fo long as they were inferior to the Gauls or Cymberi, the latter having fent colonies into Germany, as far as the Baltic fea, from Gallia Comata, the grand nurfery of Europe. Teyrn,

on the covering water one. Hence Teach and Teague, the side water confines speed, or Troat spring action; Teale, on the side water or water place; Tear, the side or face water; Teals, below the side water; Teat, the side water covering; Tectonic, covering on a place of shelter; Ted, the surface covering or grass, and a tawny person; Tedious, below the side spring; Teem, the surrounding water side, or the hay covering; Teens, within or between the coverings, tens, tyes, or boundaries; Teeth, the division or side covering; Tegument, covering on the surrounding parts or sides; Tall, above or beyond the covering place, the high light covering or the sky; Telescope, the high light confines or sky thing; Temper, the surrounding or covering part, either of animal feeling, or of the air, or climate of a country; Tempest, the sky covering the lower parts; Temple, a covered place; Temporal, on the surrounding circle part; Tempe, at the surrounding side part or country; Ten, the inclosing side or upper covering; Tenable, the inclosing or holding place; Tenant, a hold on the side, under or below; Tend, at or under the inclosing side; Tender, the inclosing water part; Tendon, under the upper covering; Tenet, holding at; Tennis, on the covering side; Tenour, the circle holding or its contents; Tense, below the inclosing side; Tent and Tenth, the covering part; Tenure, the holding usage; Tepid, at the sky part; Terce, the side water part confines; Term, the surrounding water side or country division; Ternery, the country inclosing water part of three parts or sides; Terrace, on the water part side; Terrible, on the side water place; Tertian, on the third division; Tessellated, a covering on the lower place side or ground part; Test, the lower covering side or boundary; Testate, covering the estate; Testify, the lower covering side part or a covering; Tete and Tether, a covering part; Tew, the covering side or boundary spring; Text, the outside covering; Texture, the outside covering also.

Tharlton, Herts, on the covering water place dwelling place. Thatchup, York, the covering or boundary place water gate part. ... the inclosing water side or station. Thelwall, ... lower lesser flowing water place. Thell or Thill ... water inclosing side or water place, side way, or road. ... Herts, the covering or border water place side. Thames, the covering of the water or Severn chief channel. ... on the outer parts, the King of the water confines, the founding, ... Lucius's father, the covering spring part, also ... in Wales, or Kimbaterin, Cunobelin being the chief of the inclosing water place or the

Thames. Thet and Thetford, Norf. the side or covering water part, and its way or ford. Thinne, on the covering. Thirleston, Scot. below the side water place. Thirewall, N-hum. the side spring water place or wall. Thistleworth, Midd. the side inclosing water place island side. Thoke, Norf. the covering or border water chief. Thone, Som. the border or side inclosing spring. Thoncaster or Thongcastle, Norf. Salop, a castle on the border or inclosing spring water place. Thor, on the border water man. Thorgate, the border water gate. Thoresby, by the border water side. Thornbury, Gloc. a burrow on the border water. Thornlaw, on the spring water place border. Thornton, Yorks. Bucks, a town on the border water. Thorp, the border water gate part. Thrace, the lower border or country. Thrapston or Thropston, N-ham. the border water part gate or town. Thrusk, Yorks. the side spring lower water. Thule, Thyle, or Thole, the side or border water place, and in a secondary sense a dark place. Thurkill, a chief on the water border. Thurland, Lanc. the border water part, bank side or country. Thurles, Ire. the border spring water place side or lower place. Thurso, Scot. the sea water border. Thyrn, the side inclosing water. Hence Than and Thane, on the covering, surrounding, or preferred part; Thank, a thane action; That, the part or thing, or the covering or sky part; Thatch, the covering of an inclosed place; Thaw, the surrounding spring; The, a part, side or division of a man or a thing; Theatre, the covering or boundary water part side; Thee, the or side I or man; Theft and Thief, from the side, house or possessions; Their, the sides or a man's property; Them, the side surrounding men; Theme, on the surrounding coast; Then, the in or the part or side in; Thence, below the part in; There, the part upon the covering; Thereabout, about the part upon; Thereat, at the part upon; Therefore, the part upon before; Thereupon, upon the covering part; Therewithal, the part upon or covering part on the spring side; These, the lower sides or things on the surface; They, the side men; Thick, the side water confines; Thigh, the water sides; Thill, on the sides; Thimble, surrounding the side place; Thine, man's inclosed side, possession, property, or quality; Think, a man's internal or inside action; Thing, an inclosing side, a part, or what is, exists, is possessed, or is contained within the terrestial or any particular limits; Third, the side water parts or divisions, being the water and its sides; Thirl or Thrill, the side water place, which pierces or perforates; Thirst, below the water side part; Thirteen, three on ten, tie or inclosing sides; Thirty, three water ties;

This,

This, the fide; Thiftle, covering the lower place or furface;
Thither, on the fide part; Thitherto, to the fide part or border; Though, the furrounding border or circle; Thong, the
furrounding or inclofing border, as an ifland or continent;
Thoral, on the covering part or border; Thorn, on the furrounding border by way of fence; Thorough, the border water gate; Thofe, the border fide; Thou, the fide man;
Thought, man's acting property; Thoufand, the lower border part, or fand; Thowl, the border fpring water place;
Thrall, on the fide water place; Thraldom, on the fide water
place dome; Thrafh, covering the lower or feed part; Thread,
covering the fides; Threat, at the covering water fide;
Three, the fide water; Thrice, three fides; Thrift or Thrive,
covering or inclofing the fide part or boundary; Throat, the
furrounding gate, or through the fide; Throb, the furrounding fpring part; Three, the furrounding water place or gate;
Throne, on the furrounding water fide or country; Throng,
in the furrounding border; Through, the furrounding fide
water gate; Throw, the furrounding fide water fpring; Thrum,
the fide fpring water part dumb, mute or dead; Thruft, the fide
fpring water part to the lower fide, Thumb, and Thump, the
fide fpring part; Thunder, in the hi.h fide fpring part; Thuri-
ferous, the furrounding fpring parts; Thurfday, the furrounding
fpring water gate, or Thor divifion or day; Thus, the lower
fide or face of the earth fprings of all forts; Thwack, the fide
fpring action; Thwart, at the fide fpring water; Thy, the
man's fide or part; Thyme, the fur-rounding parts or fides.

Tibury, Hants, the fide fpring water part or burrow.
Tickenhall, Worc. the inclofing water place fide, high water place or hall. Tichborn and Tichfield, Hants, the upper
fide neighbourhood or dwelling place. Tichmerfh, the water
fide marfh or meres. Tilaus, Wales, the fide fpring water
place fide. Tilbury, Hants, on the fide fpring water part.
Till and Tillerfley, N-hum. Lanc. on the fide water place
or lower thereon. Tillingham, Eff. a vale or home on the
inclofing water place fide. Tindale, N-hum. the Tine vale.
Tiningham, Scot. a vale on the inclofing water fide. Tin-
tagil, Corn. the coaft on the inclofin water place fide. Tin-
tern, Ire. on the inclofing water part fide. Tio-v ingacetter,
Nott. the fide fpring water place edge caftle, city. Ti-
perary, Ire. on the water part fide country. T, N-hum.
on the upper fide part. Tir or Tira, Wales, water
fides, divifions, parts, or poffeffions. T e, the
fpring water place fides, or poffeffions. T e, Don-
regal, Ire. the inclofing water place fides or the
fifh,

English, Welsh, or Gaulish borders possessions. Tir Oen or Owen, Ire. the water or spring water circle possessions or country. Tirvolac, Totidan, or Ireland, the inclosing water place country, or a country on the lower water border, or Danish or British possessions, and the water land. Tir-whit, the spring side country. Tisdale, the lower side vale. Tiverton, Devon, the side or division spring town. Tixal, Staff. on the side water. Hence Tiara, on the sides, possessions, or country; Tice, the lower or female side; Tick, covering the side; Ticket, at the side water confines; Tid, Tiddle, Tide, Tidily, Tidings, and Tidy, on the side water place; Tie, the side water, which both divides and ties countries, as well as persons and nations; Tier, on the side or row; Tiff, from the side; Tiger, the border water side or forest; Tight, the inclosing water side part; Tike, covering or keeping the sides or possessions; Tile, the covering place; Till, the side spring water place, or whilst it flows; Tillage, on the side water place edge, which at first was digging; Tilt, the side water place covering; Timber, the side surrounding part; Time, the part or side surrounding; Timid, at the surrounding side; Tin, on the side or covering; Tip, the side part; Tippet, at the side part; Tipple, at the water place side; Tipstaff, the side part staff; Tire, on the side; Tit, at the side; Tithe, the side or tenth share or part of the prince or governor, who at first were Druids or patriarchs; Title, the side water place, meres, or boundaries, as those of Jacob and Laban, which with possession and general consent, on taking possession of vacant countries were the original title, though right seems to be derived from the will of the Creator and supreme disposer of all good things.

Tocester, Toucester, or Treponte, Bucks, the border spring city or surrounding water parts, districts, or wapontaches. Todington, Goc. a town on the water border side. Toghs, Ire. water inclosing confines or islands, as that of the Iceni or East Angles and Trinobantes, formed by the sea and Thames. Togodumnus, the border chieftain on the surrounding spring part or the Thames, of the Iceni or Trinobantes. Tokenham, Wilts, a vale on the inclosing water side. Toliatis, Geniad or Shepey island, Kent, at the covering water place side lower part or island, on the inclosing water place side, or the island or lower part. Tollevilla, Yorks. the border or covering spring water place or village, or the toll place village or dwelling. Tong, Toang, and Tongley, Salop, the Angles, English, or inclosing water place border. Topcliff, Yorks. the side water place border part. Topsham, Devon,

Devon, the border water part side vale Torbay, Devon, the covering or border water part or bay. Torbeck, Lanc. the covering water part brooke. Toriland, Devon, the border water part bank side. Torksey, Linc. the border water side. Torleton, Gloc the bo der water place town. Torpichen, Scot. on the upper border water part. Torpul, Suff. the border water part spring p'ace. Torington, Devon, the border inclosing water town. Totnes, Devon, the promontory on the border water side. Totteshall, Linc. the covering or border water place side. Touasis or Tees, Dur. the lower side or border spring. Touchet, Chef. the border spring water confines side. Tou, Tovi, or Dovi, Devon, Wales, the border spring. Toam, Ire. the surrounding water border. Towridge, Devon, the border spring water edge, way, ford, or gate. Towton, Yorks. the covering or border spring water town. Hence To, the surrounding side, covering, border or side from ; Toad, at the water side ; Toast, at the lower side ; Tod, the covering part; Toe, the side covering or edge ; Toft, the border part side or house ; Together, on the surrounding water side or country ; Toil, on the surrounding water place ; Token, on the surrounding water confines, where beacons were placed, as that of Laban and Jacob, and those along the sea coasts ; Told, Toll, Tolerate, Tolboth, and Tollsey, on, at, or at the side of the border water place, the toll place, the boundary river toll part, the sea toll place, and the toled or told place ; Tomb and Tome, or the surrounding or covering part ; Ton, Tone, Tong, and Tonnage, the internal part, or contents of the water lower Ton, a buckle ring or a wave; Tongue, the wave of the founding spring, or the spring in the mouth or surrounding water or sea part ; Too, over or from the side or border ; Tool, the border or side spring water place ; Tooth, the covering spring side ; Top, the covering part, as the sky or other covering ; Toparch, the border part chieftain ; Toper, the water part covering ; Tophet, the lower confines part or hell ; Topick, the boundary part confines or action ; Topping, the inclosing water part covering, or border ; Topsyturvy, the border or covering part on the side spring ; Tor, the water covering, tower, or belly ; Torch, the upper or light border or covering ; Torment, at the border or tower part ; Tormentor, the border part or torment man ; Torpent, Torpid, Torpor, and Torrent, on the covering part ; Tort, at the side border ; Toss, the water border side or waves; Total, on the border side possessions, or country ; Touch, the upper or spring covering; Tough, the inclosing spring water border ; Tour, on the spring water border ; Tow, the spring covering;

Towards,

Towards, on the spring side; Town, on the border spring covering, or inclosure on the spring; Toy, the covering spring.

TRAETH, Wales, the water part side, sands, or lower water part. Trafford, Lanc. the sand way or passage. Trajectus or Oldbury, Gloc. the water part confines side or passage, or the circle water place side spring part or burrow, and an old burrow in a secondary sense, as ancient dwellings were situated on the confines spring or water sides. Trailey, Ire. the side water part p'ace, or the ebbing water place. Tralwin or Welshpool, N-wales, on the covering water place spring, or the Welsh furrounding water place, or passage. Trawisfyneth, N-Wales, the spring water part side, or passage mountain. Hence Trace, Track, and Tract, the water part side, street, or town; Trade, at the water part; Traffic, the water part confines or action part; Trail, the water part or passage place; Train, on the water part; Traipse, the water part low or side part; Traitor, from the side water part or border; Trammel, the part furrounding the water place; Trance, on the water part side; Transact, acting on the water part side or passage; Transfer, on the water part side or passage part; Transfuse, the side spring in the covering water part; Transport, over the furrounding water part side; Trap and Trappings, covering or furrounding parts or things; Trash, the lower side or surface; Travel, the spring water place side; Traverse, the water part at the side or below the spring; Tray, the covering or side water part.

TREBORTH or Treboeth, Chef. Wales, the water part side or town port, gate or entrance. And as it was usual on attacks to set the suburbs on fire, the burnt or burning town. Tredenham, Corn. a vale or dwelling place below the water part side, street or town. Tredanoe, S Wales, a town below the water side. Trefusis, Corn. a town below the side spring or spring side, or a smelting town. Tregaren, S-Wales, a town on the incloling water part or border. Tredorman, Cumb. on the furrounding water part side town. Tregeney, Tregonen, Tregonwell, Tregoze, Tregna, Trelech and Trelauny, Corn. Devon, Wilts, Wales, a town on the incloling water or spring water side, bank or lock. Trent, Dorset, Nott. Staff. the side incloling water part; Trentham, Staff. the Trent vale. Treeves, of Belgic Gaul, the side spring water part side, street or town. Hence Tread, at the side water part; Treafon, on the water part lower side or below the border; Treafure, on the spring water part side; Treat and Treaty, at the side water parts; Tree, on the water part side; Tremour, the side

furrounding

furrounding water part or the fea; Trench, the inclofing water part confines or ditch; Trepan, on the fide water part; Trefpafs, paffing below the fide water part; Trefs, below the fide water part; Trefpaffing, paffing the fame; Tret, at the fide water part; Trevet, three fides; Trey, the three or fide fpring water parts confifting of the water and the two fides, banks or poffeffions.

TRIG or Trige, Dorfet, N-Wales, the fide water confines, habitations or poffeffions. Trim, Ire. the middle or border water part. Trimlets Town, Ire. the divifion or furrounding water place fide town. Trimontium, in Thrace, the three or furrounding water part, fides or mountains. Trinobantes, Midd. Eff. the fide water part or town in the vale or bottom or Trinoventes, the town on the fpring fide or of the Wentas. Tripoly, the furrounding water place town. Trifanton or Southampton, the lower town on the water fide. Hence Triad, three fides, divifions or parts; Trial, three calls of O-yes or the circle attendance at the fide or borders, where courts were held; Triangle, three angles or corners; Tribe, the fide water part or race; Tribune, the tribe one or on the fide water part; Tribute, the tribe ufage; Trice, Trick and Trickle, the fide water confines or action; Trifler, the fide rifler; Trill, the rill fide; Trim, the furrounding water part fide; Trinity, three united; Trip, the fide part up; Trite, at the fide part; Trivial, on the fpring water fide; Triumph, the furrounding fpring water parts or fhouts of joy, which were three exclamations ufually made on the borders on the defeat of affailants.

TROI, the furrounding or weftern water fide or town. Trotman, the border water or fea fide part haven or man. Trovis, Ire. the furrounding fpring water fide, parts or poffeffions. Trompette Caftle, in Gafcony, a caftle at the fide furrounding water part. Hence Troat, at the fide furround ng water part; Troll, on the fide furrounding or circle water place or its roll; Troop, the fide furrounding part; Troth, at the circle water part fide or oath; Trouble, the fide furrounding fpring water place; Trover, over the circle water part; Trout, at the furrounding fpring water part; Troy, the circle or border water fide.

TRUBRIDGE, Wilts, the fide fpring water part edge, bridge, ford or way Truardraeth, Corn. the fide fpring water on the fands. True Place, N-hum. the fide fpring water place or paffage. Trumpington, Camb. a town on the furrounding fpring water part fide. Truro and Trury, Corn. Bedf. on the fide fpring water part, paffage or town. Hence Truant, fide

4   of

of the spring water part or the side spring wanting; Truce, the side spring water inclosed, lowered or shut; Truck, the side spring water confines or action; Trudge, the side spring water edge or action; True or Truth, the side or surrounding spring; Trull, on the side spring water place; Trumpet, the side surrounding spring or hollow part, Trundle, on the inclosing spring water place side; Trust, at the side spring water part side; Try, at the side water part.

Tuchet, Wilts, at the side spring water confines. Tuchwic, Oxon. the side spring water confines, fortification or street. Tudenham, Gloc. the vale or forest below the side spring water. Tuddington, Bedf. a town on or below the side spring water part. Tuisco or Tywisog, the side spring water confines leader, chieftain or prince. Tulibardin, Scot. on the side spring water place part side or the bard family or tribe. Tulket, Lanc. at the side spring water place confines. Tumuli, the surrounding spring water place sides, tombs or barrows. Tunbridge, Kent, a bridge on the side spring water. Tungri, Cumb. on the side spring water confines part. Tunocellum or Tinmouth, N-hum. the side spring water on the surrounding or salt water place or the Tine mouth. Turberville, the side spring water part dwelling place or village. Turbery, the side spring water part or boggy places left undrained for the defence of the borders. Turditania, Spain, below the side spring water part. Turlogh, Ire. the side spring water lake. Turman or Truman, the side spring water part. Turthill Fields, Midd. fields at the side spring water part or dunghill fields. Turton, Lanc. the side spring water border or town. Tutesbury, Staff. a burrow below the side spring water. Tuxford, Nott. the side spring water confines way or ford. Tweede, N-hum. the side spring water part. Tweedesdale, the Tweede side vale. Twist, the side spring water lower side. Turford, N-hum. the side spring water way or ford. Tuomond, Ire. on the surrounding spring water side. Hence Tub and Tube, the side spring water part; Tuck, the side spring water confines; Tuesday, the side spring's day or division; Tuft, at the side spring part; Tug, the side spring water confines or action; Tuition, on the side spring side; Tulip, the spring water place sides or two lips or parts; Tumble, the surrounding spring water place side; Tumefy, the surrounding spring part; Tump, the surrounding spring side part; Tumult, at the surrounding spring water place side; Tun and Tune, on the side spring; Tunick, on the side spring covering; Tup, the side spring part; Turbary, the side spring water part or common; Turbulent, the side spring water part bank; Turd and Turf,

at

at the spring water part or side; Turmoil, the surrounding spring
water place; Turn or Tourn, on the border spring water part or
the earl's or sheriff's tourn in the county; Turret, at the side spring
water part; Tush, below or the lower side spring; Tutelar,
on the side spring water place side; Tutor, at the side spring
man; Tux, the side spring water confines; Twain and Tway,
the sides of the inclosing spring water; Tweedle, at the side
spring water place; Twelfth and Twelve, two above the side
part or border; Tye, the sides or ten parts; Twenty, two tyes
in or tens; Twig, the side spring growth; Twilight, two
lights; Twin, two in; Twirl, the side spring water place;
Twixt, the sides of the inclosing spring water. Two, the
surrounding spring water sides, un or one being the spring,
and three the inclosing spring and its sides; Tye, the side poffeffion or property; Type, the tye part, letter or emblem of a
thing; Tyranny, the side water border or country divisions, and
in a secondary sense, a cruel lord, governor or prince or an
usurper; Tylwith, the side spring water place side, house, poffeffion, tribe or family.

VACCARII, Bercarii or Hull, Yorkf. the spring water or
water part confines, borders or cities. Vacomagi or Murrey,
Scot. the surrounding spring water confines or vale or vale nation. Vadimon, Scot. the surrounding spring water side
mountain or Lomond lake. Vaenor, Wales, the inclosing spring
water border or a vale. Vagniaccæ or Maidston, Kent, on the
inclosing or side spring water confines or town. Vale or Fale,
Corn. a spring water place or a vale. Vale or Falmouth, its
mouth. Vallum, a surrounding spring water place, a ditch
or a wall. Valentia, on the spring water place bank or wall
side or the name of the Roman province on Hadrian's wall.
Valesii, the spring water place or vale sides or the Welsh or
Gauls. Valtort, Corn. Devon, the spring water place border
side. Valvafor, Yorkf. the spring water place side man, servant
or vaffal. Vandal or Wental, Surry, the Wenta or the inclosing spring water place coasts or vales. Vandalbiria, Camb.
the Wenta, Veneti or Vandal water parts or the spring coast-
ing water places; the Wenta, Veneti, Venedi and Vandales
being only different names of the same people of Gaul, Britain,
Italy, Mecklenburgh, Brandenburgh, Pomerania, Livonia,
Lithuania, and other parts of Poland, Germany and Denmark, who were very confiderable tribes of the Cumberi of
Gallia Comata, and Bracata, who from their dwelling places
and employments of draining and improving countries,
along the river sides or vales, acquired that name, and thence

the

the Eng'ifh name Wander. Vararis or Murray and Varis, Scot. N-Wales, the incloſing ſpring water or ſea ſide or a marſh. Varini or Werini, of Mccklenburgh and Pomerania, on the incloſing ſpring water ſide inhabitants or the Veneti Vaſcones, Spain, the lower water confincs. Vaſates, of Gaul, at the water ſides. Hence Vacant, the ſpring water incloſing ſides ; Vagabond, from the ſpring water boundary ; Vague, the ſpring water confines ; Vail and Vale, the ſpring water place or covering or incloſing a place in a ſecondary ſenſe ; Vain, the covering upon ; Valetudinarian, one that has a covering on his ſides ; Valiant and Valour, on the ſide ſpring water place ; Valid, the ſide ſpring water place; Value, on the ſpring water place ; Valve, the ſpring water place part ; Vamp, the ſurrounding ſpring or covering part ; Van, the incloſing ſpring, which is in the front ; Vane, Vann and Fan, in or the incloſing river or haven part ; Vaniſh, lowering the vann or part ; Vanquiſh, the vane or colours down ; Vantage, the ſpring incloſing or Veneti coaſts ; Vapid and Vapour, the ſpring part ſide ; Vary, the ſpring ; Vaſe, the ſpring ſide ; Vaſſal, on the ſide ſpring or an upper ſervant ; Vaſt, the ſpring ſide coaſt ; Vat, the ſpring ſide ; Vavaſour, the ſide ſpring water man, ſervant or vaſſal ; Vault, on the ſide ſpring water place ; Vaunt and Vaward, the ſpring part on the ſide or the firſt or fore part.

Ubranford, N-hum. the upper incloſing ſpring water ford or way. Ubartus, a river of Italy, the ſide ſpring water part. Ubii or Cologne, in Germany, the incloſing ſpring water part, bank or lane. Ublogahel, Ire. the Gauliſh ſpring water place. Ucheltre, Scot. N-Wales, on the upper incloſing ſpring water part or hills. Udeceſter, the ſpring ſide city. Hence Uberty, the ſpring water part ſide ; Udder, the ſide ſpring water part of an animal ; Ubiquitary, on the incloſing ſpring water part confines or wandering every where in ſtreets on the river ſides.

Vecta or Vectis, Hants, the incloſing water ſide, iſland, or lower part or poſſeſſions. Vecturiones, Scot. the lower or weſtern incloſing water ſide, iſlands, or iſlanders. Vedra, Virus, or Wore, Dur. the ſpring water part. Velabri, Ire. the ſpring water place or vale water parts. Iberi, the water parts or Britons. Vellocatus, the armour bearer of Venutius, the guardian of the rivers. Veneti, Wenta, or Went, the ſpring coaſts of the Iceni or the iſland water confines, or the incloſing water place confines, Belgæ or the incloſing water place confines, and Silures, the lower ſide ſpring water place ſide. Veniticæ, Venis, or Venice, Italy, the incloſing water ſide, iſlands, or lower parts. Venutius,

tius, king of the Venedotians, and husband of Cartismandua, the Veneti, or the inclosing spring coasts or vale possessions or possessors. Vennicnium, Ire. an inclosing spring water confines promontory. Verbeia, Wheref, Yorkf. the spring water part. Verdon, or Ireland, in the border water, or the water land. Vergivian, or Werith sea, the side or lower water. Veslucio, or Werminster, Wilts, the side spring water place confines or monastery. Vernicones, or Mernis, Scot. the inclosing water or sea confines or islands. Vernolium, the inclosing spring water place. Vernometum, or Burghill, Leic. on the surrounding spring side. Verulanum, Urolanzum, Wætlingaceaster, or Saint Albans, Herts, the surrounding or side spring water place, bank, street, or city. Verteræ or Bury, under Stanmore, W-mor. the spring water on the border water, or sea side. Ufford, Suff. the spring part or upper ford or way. Hence Veal, the spring water or meadow place, or a meal; Vection, on the side spring water; Veer, on the spring; Vegetable, on the springing at the side part; Vehement, the high spring part; Vein, the inclosed spring; Velocity, the spring water place; Venal, the inclosing spring water place; Vend, the inclosing spring water side; Venerable, able on the inclosing spring water; Venery, on the inclosing spring water; Venison, the inclosing spring side one; Vent, the inclosing spring side; Verb, the spring or water spring part; Verberate, at the spring water part : Verdant and Verdure, below the water spring side; Verdict, the spring side act, or an act of truth; Verge, the inclosing water confines; Verify, the spring water part mark or border; Vernacular, together on the spring water place confines; Vernal, in the spring; Verse, the spring side; Vert, a spring part; Very, the spring; Verity, the spring tie, division, side or part; Vesicle, the side spring shut place; Vesper, the spring lower part; Vessel, on the spring water place; Vest, and Vesture, lowering the lesser spring or virgin; Vestry, on the spring side or fore part inclosure; Veteran, the side spring water one; Vex, the spring water confines action; Ugly or Ogley, the surrounding spring or sea water place.

Vicus Malbanus, or Nantwich, Chef. a street on the middle water part, or the mall. Vidogara, Scot. the covering spring water confines. Vidua, Ire. the side spring. Vignones, Germany, on the spring water confines or streets. Villa, on the spring water or dwelling place. Villa Faustini, or Edmonsbury, Suff. a dwelling place or village on the spring side part. Vindogladia or Winbourn, Dorset, on the spring water border, country or neighbourhood. Vindolana, N-hum. on the spring

water

water border banks. Vindobala or Wallefend, N-hum. on the border fpring water place or vale fide. Vindomora N hum. on the furrounding water or fea fide or edge. Vindonum or Cynegium, Ventenfe, Bentenfe, Winton, or Wenta, Hants, on the inclofing fpring border or Kentifh confines town. Vinderius, Ire. the inclofing fpring water part. Vinovium, Binovium, or Binchefter, Dur. dwellings or a city on the inclofing fpring water edge. Vipont. on the furrounding fpring water part. Vipfeys, Yorkf. the fpring parts from below. Viroconovium, S-Wales, the inclofing fpring water or dwelling place. Virofidum, Cumb. the fpring water fide. Virvedrum or Dunfby. Scot. the part or dwelling on the fpring water fide. Vifi Gothi, the fpring water fide Goths. Vifi Saxones, the fpring water fide Saxons, the Goths and Saxons being moftly fea coafters. Vitæ or Witæ, Hants, the fide water coafts and coafters. Vitfan or Whitfan in France, on the lower water fide. Hence Vial, on the fpring water or food ; Viand, the inclofing fpring fide ; Vibrate, at the fpring water part ; Vicar, on the inclofing fpring water border ftreet or village ; Vice, a leffer or fecond on the fpring fide or below the fame ; Vicinity, the inclofing fpring water confines fide ; Victim, on the furrounding fpring water confines ; Victor, the fpring water border man ; Victualler, on the inclofing fpring water place man ; Vie, the fpring ; View, the fpring of man or animal ; Vigil, on the inclofing fpring water fide ; Vigour, man's fpring ; Vile, the fpring water or dwelling place ; Villain, on the fpring water place bank, village, or dwelling place ; Vindicate, keeping the inclofing fpring fide ; Vine, the fpring one ; Violate, at the furrounding fpring water place ; Viper, the fpringing or ftinging part or thing ; Virgin, in the fpring or beginning ; Virtue, the fpring fide ; Virus, fpringing from below ; Vifage. Vifit, the fide or fight fpring ; Vital, on the fpring fide ; Vivid, a lively fpring ; Vixen, the inclofing or fhut fpring water confines one.

ULCOMB, Kent, the fpring water place vale. Uleigh, Gloc. the inclofing fpring water place. Uleftanton, Salop, below the fide fpring water place confines or town. Uliarus, the fide or lower water place country, or an ifland. Ullefwater, W-mor. the fide fpring water place. Ulmetum, Yorkf. the furrounding fpring water place fide. Ulfter, Ire. the fpring water place fide part. Hence Ulcer, fpringing place on the fide ; Ultimate, at the furrounding fpring water place ; Ultramarine, the furrounding fpring water place or fea fide.

UMFRANVILLE, Hert. Salop, N-hum. a village on the furrounding fpring water part. Umbria, or Chumbria, Italy, the

furrounding

furrounding fpring water part or vale country. Hence Um-
bles, the furrounding or border fpring water places; Umbrage,
the furrounding fpring water part edge or action; Umpire, over
or on the furrounding fpring water part

UNESLAW, Herts, the incloſing fpring fide fpring water
place. Una or Sus, a river of Mauritania, the fide or incloſing
fpring. Undalum, Gaul, the incloſing fpring fide water place
or vale. Hence Unckle, one incloſing fpring water place or
of the fame country or family; Unction, on the incloſing
fpring fide; under, the incloſing fpring water part; under-
ſtand, the incloſing fpring water part ſtanding; Undertake,
the incloſing fpring water part coaſt keeping; Unicorn, one
horn; Union, Unit, Unite, and Unity, the incloſing fpring
water which was but one; Untⁱl or untie, the incloſing fpring
border or covering, and unprefixed, unable, unactive, unfelt,
unfit, unguard, unlace, ungird, untangle, unwife, and various
other words, as being ſhut up or an incloſed fpring fignify
a negative or privative.

VODIÆ or Oudiæ, Ire. the furrounding fpring fides or pof-
feſſions. Voelas, Wales, the furrounding fpring water place
fide or green hill. Volantium, Cumb. the furrounding fpring
bank fide. Voluba, Corn. the furrounding fpring water place
part or vale. Hence Vocal or the furrounding fpring water
place, cheeks or mouth; Vocation, on the furrounding fpring
water fide; Vogue, the furrounding fpring confines; Void,
the furrounding fpring fide; Voice, the furrounding fpring of
found; Vollant, the fpring on the high fide; Volition, the
light divifion fpring; Volley, on the fpring water place; Vo-
luptuous, the furrounding fpring water place or belly fide;
Vomit, the furrounding fide fpringing up; Vortex, the incloſing or border fpring water confines; Vote, at the furrounding fpring water; Vouch, the furrounding fpring water
confines or act; Vow, the furrounding fpring oath or woe;
Vowel, on the furrounding fpring water place; Voyage,
the furrounding fpring or water edge or action.

UPHAVEN, Wilts, the Avon or river fpring or upper part.
Uppingham, Rut. the incloſing fpring or upper water part or
vale. Upton, Worc. N-ham. on the border fpring part, or
the fpring part town. Hence Up, the fpring part, which is its
upper part; Upon, the fpring part on; Upper, the fpring
water part; Uppermoſt, the fpring water part furrounding the
lower fide; Upfide, the fpring part fide or bank, which is the
upper part; Upward, the fpring at the fide water which rifes
up from below, and as it increafes in the main river grows
higher.

URE

URE, Yorkf. the fpring water. Urefby, Linc. the fpring water fide part or habitation. Uriconium, Wreeken, or Wroxefter, Salop, the inclofing fpring water part. Urnfton, Lanc. the furrounding fpring fide town. Hence Urbanity, the city part; Urchin, fp in ing one; Urge, the fp ing edge; Urine, the inner or inclofing fpring water or in man; Urn, the fpring water, or man's inclofure or fnut

USHANT, on the lower fide Ufk, Wyfk, or Burrium, S-Wales, the lower or fide fpring water. Ufocona or Okenyate, Salop, the lower or inclofing fpring fide or gate. Ufipii, Germany, the lower or fide fpring parts. Hence Us, the fprings of men; Ufage, the actions of thofe fprings or men; Ufe, the fpring fide; Ufurp, on the fpring water fide part; Ufury, the ufe men.

URBORROW, the fpring fide burrow. Utocetum or Utokcefter, Staff. the inclofing fpring water fide border or city. Uthred, at the fpring fide part. Vnedal and Vnfald, the fpring fide water place or vale part. Uxbridge, Midd the inclofing fpring water confines bridge. Uxellodunum, Uxella, or Ledfithic, Corn. Scot. the inclofing fpring water place fide or hill Vyrnwy or Vrynwy, Wales, the inclofing fpring water part or hill. Uxellum or Fufcr, Scot. the fide or lower fpring water place. Hence Utenfil, below the fpring water place fide; Uterine, the inclofing fpring water part; Utility, the fpring water place fide, poffeffions, or property; Utmoft, the fide or lower furrounding water parts or coalts; Utter, the fide fpring part; Vulgar, dwellers in villages, or on the inclofing fpring water places; Uxorious, the inclofing fpring water confines man.

WABRIDGE, Wadbridge, or Warbridge. Corn. Hunt. the fide fpring water bridge. Wachopdale, Scot. the inclofing fpring water part vale. Wahul, Odil, or Woodhill. Bedf. the furrounding fpring water fide, high place, or hill Wainfleet, Linc. the inclofing fpring flood fide. Wakefield, Yorkf. the fpring water confines dwelling place. Walbrook, Midd. the fpring water place or wall brook. Walcot, Linc. the fpring water place coat. Waldgrave, the fpring water place fide warden. Wale, Wales, Vales, Waley, in Weft Britain, Yorkf. Lanc. Germany, fignify, the vales or fpring or lower water place fides, a Saxon tranflation of Cumbri, the vale Britons, but Wales, till king Ina's time, from Beri in Cumbri, went by the name of Brittenmen, the Saxons, being confidered as diftinct people, as Englifhmen and feamen. The prefent Welfh or Cambri Britons, who among themfelves go by the name of Cambri, of whom the illiterate know

not

not to this day that they go by any other name in other countries, are probably Gaulifh or Englifh colonies, grafted on the
Aboriginal Britons, according to the very import of the name
Cumberi, and their names of perfons and places; though their
paffage over the Englifh channel, as well as the other inhabitants of the weftern parts, might perhaps have been fomewhat
fooner than of thofe on the eaft fide of the ifland; and whatever
difference there may now appear betwixt their dialects in the
compound, they ftill appear to be in the abftract the fame, and
to be of thofe dialects, which had their rife chiefly in this
ifland, as appears by the principles of their conftruction, and
names of perfons and places. And though fome hiftorians
pretend that the Welfh were expelled England by the Saxons,
there are provifions made by the laws of the feveral Saxon hep-
tarchies, for the protection of the life of Welfhmen or landmen, as well as Saxons or feamen, and confequently for their
protection in the country, though the fines differ in value, as
the Saxons or feamen, who confidered themfelves as being derived of the Britifh or Irifh origin, regarded the Welfh or
Cumberi as ftrangers to them in point of origin. Walefbo-
rough, Waleton, Walford, Walflect, and Walland, Corn.
Suff. Der. Lanc. Herts, Eff. Kent, a burrow town, ford,
flood and land or bank fide on the fpring water place. Wallingford, Galleva Attrebatum, Calleva or Galleva, Berks,
Oxen. the inclofing fpring water place, ford, or way of the
Attrebati, or the divifion water part fide. Walmer, Kent,
the fea water place. Walsford, Hunt. N-ham. the furrounding fpring water place fide, way, or ford. Walney, Lanc.
the inclofed water place. Wallup, the fpring water place part.
Walpole, Norf. the fpring water place, pool, or lake. Wal-
fingham, Norf. the inclofing water place fide, vale, or houfe.
Walfal, Staff. on the fide or lower fpring water place. Wal-
tham, Eff. the fide or lower fpring water place, vale, foreft,
or wood. Walwick, N-ham, the fpring water place or wall,
ftreet or fortification. Wanborough, Wandlefbury, Wandf-
worth, Wanfditch, Wanfted, Wantage, and Wantfum or
Stour, Wilts, Camb. Surry, Eff. Berks, Norf. Kent, a burrow, a being part, a ditch, a ftation, an edge and vale on the
inclofing fpring water fide, or the Wenta or Veneti borders.
Wapentack, Wapentache, or Pontes, the inclofing fpring water part coafts, or certain diftricts of Nott. Yorkf. Linc. Leic.
N ham. Der. and other northern counties of the fame fignification, as hundreds or chantels in the more fouthern parts,
which confifted of the Wenta and Brigantes of Ponticu, or Pliny's
Britania or Brigantes of Gaul. Walden or Wardon, N-ham.

on the spring water side. Warbour, Wilts, the spring water part gate or tower. Ware, Weare, Were, or Vedra, War. Wilts, Herts, Dur. the spring water or spring water part. Warham, Dorset, the spring water vale or home. Wark or Werk, Som. N-hum. the spring water confines. Warkworth, Dur. N-hum. by the spring water confines. Warmington, War. the surrounding spring water confines town. Warnford, Hants, the inclosing spring water way or ford. Warington, Lanc. the inclosing spring water town. Warwick or War-in-wic, the inclosing spring water street or fortification. Washes, Rut. the low springs. Washborn, Worc. the low springs neighbourhood. Waterford, Ire. the spring water side, or part way or ford. Watford, Herts, the side spring way or ford. Watlin-street, from Dover to Anglesey, by London, St. Albans, Dunstable, Stony Stratford, Toucester, Littleborn, or St Gilbert's Hill, &c. the spring water bank, street, or the place of the migration of the clan, land, or Gaulish race. Watchbury, Salop, the water place side burrow. Watlington, Salop, the inclosing spring water confines, side, or bank town, or Watling-street town. Wauburn, Norf. on the spring water part. Waveney, Suff. the spring water part. Waverley. Surry, the spring water place. Hence Wabble, the water place spring part, or wave; Wad, the side spring; Waddle, Wade, Waft, Wage, and Waggith, the side spring water part, place, side, or confines; Waggon, on the inclosing spring water; Waif, off the inclosing or border water; Wail, the spring water place; Wain, the inclosing spring or on the spring; Wait, the spring side; Wake, the spring water confines; Walk, the spring place, or man in action; Wall, on the spring water place, or a bank of turf; Wallop, the spring water place or vale upper part; Wamble, the surrounding spring water place; Wan, spring in or shut; Wand, the inclosing spring part; Wander, on the spring water part or river sides, along which the Gauls or Cumbri made their migrations into most parts of Gaul, Germany and Britain, and from whom came the Vandals, Veneti, and Venedi; Want and Wont, the spring that in; War, on the spring; Warble, the spring water place; Ward, on the side spring; Wares and Wary, on the side spring; Warm, the surrounding spring; Warning, on the inclosing water confines; Warp, the water part; Warrant and Warranty, on the inclosing spring water side; Wart, a spring on the side; Was, the spring past; Wath, the spring side; Watle, the spring side part; Watch, at the inclosing spring water confines; Wattles, the spring water place side, or streets built of hurdles, along the

river

river fides, like Watling-ftreet; Wave, the water fpring;
Wax, the inclofing fring, or fticking together water; Way,
the water. or its confines; Waylay. the way place or on the
way; Wayward, on the fide fpring way or froward.

WEADMORE, Som. the water fide moore. Wealde or
Weildie, Kent, the fpring water place fide, wild, or wood.
Wealis, Welfh, or Gaulifh, the fpring water place, fide, or
vales, or a people fated on the weftern fide of England, who
amongft themfelves go by the name of Cymberi, or vale Britons, being colonies of the Gauls or Englifh, from the eaftern
parts of England, grafted on the Old Britons, but erroneoufly
fuppofing themfelves as a branch of the Cymberi nation, to
differ in their origin from the reft of their countrymen of England, but if they really are Cymbri, as their language, names
of places and perfons, and antiquities indicate, they muft be
entirely of the fame origin as the other Englifh, though their
paffage over the Englifh channel into Britain, as well as others
who inhabit the weftern parts of the ifland, might perhaps have
been a little fooner than the eafterns. Weeft or Weft Mean,
Hants, the middle or divifion water lower fide. Weble Here.
a place without a fpring. Wedfbury, Staff. the lower fpring
water or river fide, burrow or dwelling place. Weden on the
ftreet, or Bona Venta, N-ham. the fpring water town on the
ftreet, or Watling ftreet, or the inclofing fpring or Veneti
parts. from whence North Wales had the Roman denomination
of Venedotia, the Veneti boundaries. Weels, on the water
places. Weeting, Norf. the inclofing water fide. Welsford,
Ire. the water fide or lower way or ford. Welbee, Nott. the
fpring water place brook, or dwelling place. Wells or Theo-
dorodunum, Som. poffeffions on the fpring water places.
Welland, Line. Lei. the fpring water place bank fide or land,
Wellington, Som. Salop, the inclofing fpring water place.
Watling ftreet. or an Englifh town. Wem and Wenfis, Salop,
Scot. the inclofing fpring and the furrounding fpring fide.
Wenlech, ——, the inclofing fpring water place fhut. Wen-
man, the inclofing fpring, Veneti or Wenta part. Wenmer
or Danco, Here, the inclofing fpring water or Veneti meeres.
Went, Wenta, or Veneti, the inclofing fpring fides, coafts,
or diftricts. Wentdale, Yerki, their vale. Wentfern, Norf.
their inclofing water. Wentworth, Yorkf. their fide fpring
water, and from the Wenti the prefent Englifh dialect in a
great meafure derives its origin. Weorth or Worth, the fpring
or river on the fide, which inclofed fmall parts or farms.
Weremouth, Dur. the fpring water or river mouth. Wermin-
fter,

ſter, Wilts, on the ſurrounding ſpring water ſide part, ſtation, or, monaſtery. Weſtlan, Devon, below the ſpring water or river ſide. Weſt and Antiveſtum, Corn. the lower water ſide and on the ſide of the lower water ſide or well, it being the weſtern boundary of the Celtes, as the Eſtii of Pruſſia and Livonia, whoſe inhabitants ſpoke the Britiſh tongue in the days of Tacitus were their eaſtern limits. Weſtmorland, the weſtern ſea bank ſide, diviſion or country. Weſtminſter, Midd. the lower water or river ſide minſter or monaſtery. Wetherby, Yorkſ. the ſpring water part or habitation. Wever and Wey, Staff Cheſ. Dorſet, and Surry, the ſpring water. Wexford, Ire. the ſpring water confines, way, or ford. Weymouth, Dorſet, the ſpring or river mouth, or at the ſurrounding water or ſea ſide. Hence We, the ſprings or men; Weak, from a ſpring; Wealth, the ſpring water place ſide, poſſeſſions, or properties; Wean, ſhutting the ſpring water ſide, or the breaſt; Weapon, on the ſpring water ſurrounding part; Wear and Weary, on the ſpring water; Weaſel, a ſmall animal; Weather, the ſpring ſide or diviſions; Weave and Web, joining together like the ſpring water parts or diſtricts; Wed, the ſide ſwings together; Wedge, the edge or ſide ſpring or ingot; Wedlock, the ſide ſpring lock; Wedneſday, be ow the ſide ſpring day or diviſion; Weed, the ſpring parts; Week, the ſpring water confines or action; Ween, in us; Weep, without a ſpring; Weevil, an evil ſpring; Weſt, off the ſpring ſide or incloſed part; Weight, at the incloſing ſpring water and ſea ſide, where the ſtrength and weight of the two waters are diſtinguiſhed, as that of the Teſt and ſea water at the Iſle of Wight, which may derive its name from thence; Welcome, the ſurrounding ſpring water, or dwelling place or commot; Weld, the ſide ſpring water place; Welfare, on the ſpring water place part; Well, a low ſpring water place; Welt, the ſide ſpring water place; Wen, a ſpring upon; Wench and Wenching, the ſpring in or meeting the incloſing water or ſea confines; Were and Wert, the ſpring at the water or ſea ſide; Werth, Wrth, and Weorth, a ſpring or dwelling on the ſide water, as Wandſworth by the Thames ſide; Weſt, the ſpring or lower water ſide; Wet, the ſpring ſide; Wether, on the ſide ſpring; Wex, ſpring out.

WHALY, and Wharlton, Wharton, or Whadon, Bucks, Devon, Lanc. the ſpring water ſide, place, or town, as are Hadon, Hatten, Hadley, &c. in which the w has been dropped, though the refinement, as in various other inſtances, has taken away the meaning, and made the names arbitrary and and indefinable. Wherp, Whert, or Verbeia, Yorkſ. the ſpring
water

water part. Whethamſted, Herts, the ſide ſpring vale, dwelling part, or ſtation. Whinfield, W-mor. the incloſing ſpring vale or dwelling place Whitby. Whitcheſter, Whitcoat, Whithern, Whitley, Whitney, Whitſand, and Whittington, Yorkſ. N-hum. Scot. W-mor. Cheſ. Cumb. Lanc. Oxon. Hereſ. Kent, and Salop, the ſide ſpring or water part or abode, city or caſtle, coaſt, incloſures or houſes, haven or ſpring part, place, ſide, or ſand Whorweil, Hants, the ſide ſpring water part on the ſea ſide. Whotſpur, N hum. the ſide ſpring water at the ſea ſide. Hence Whale, the chief or higher water animal; Wharf, the water part; What, the ſide ſpring water, which divides countries into parts; Wheat and White, at the upper ſpring, the eye or ſky; Wheedle, the ſpring at the water place; Wheel, the upper ſpring place, or the ſun; Whelm, the ſurrounding ſpring water place; Whelp, the water place ſpring part or thing; When, in the ſpring; Whence, below or at the ſide of the ſpring; Where, on the ſpring part; Wherefore, on the ſurrounding ſpring water part; Whet, the ſpring ſide; Whether, on the ſpring water ſides; Whey, the ſpring water; Which, the eye, ſky, above, or the upper ſpring or action; Whiff, the mouth or wind ſpring; While, the light ſpring; Whim, the ſurrounding ſpring water part; Whimper, the whim water part; Whin and Whine, on the ſ ring; Whip, the ſpring part; Whirl and Whirlpool, the incloſed ſpring water place or lake; Whiſk, the leſſer ſpring water confines; Whiſper, the leſſer wind ſpring part; Whiſt, at the leſſer wind ſpring, or ſilent; Whiſtle, at the leſſer wind ſpring place; White, at the ſpring; Who, the ſpring of man; Whole, the ſurrounding ſpring water place; Whom, the ſurrounding ſpring of man; Whoop, the wind ſpring up; Whore, the ſurrounding ſpring water part; Whoſe, the ſpring or man's ſide; Why, the ſpring.

Wic, Wik, Wyke, Wiccia, Wiccii, Wiccingi and Wich, of ſeveral parts of England and Ireland, the incloſing ſpring water confines, ſtreets, fortifications, villages and cities. Wicker or Boats, the boats thereof or ſuch as were uſed on the Severn. Wickham or Wic le, Wickhampton and Wicklow or Blani, Bucks, Hants, Dck , Ire. the incloſing ſpring water confines vale, water banks or dwelling places. Widevill, the ſpring ſide dwelling place. Widdington, the incloſing ſpring water confines town. Widdeham, the ſpring water part dwelling place. Widchay Berks, the high ſide ſpring. Wider, Wales, the ſpring water part. Withe, con the ſewing ſpring or ſpring water place. Wiggin or Wilwigan, Lanc. on the ſpring part,

beginning

beginning or fource. Wight, Vectis or Gwith, Hants, below or at the fide of the inclofing water at the fpring part or ifland. Wigmere, Heref. the mere fpring water. Wigton, Yorkf. Scot. the fpring water confines town. Wiift, Ire. below the water fide. Wilberham, the fpring water p'ace vale. Willey, Willeley, Willeford, the fpring water place ford or way. Willough'y, the fpring water place confines abode. Willimotes, N-hum. the furrounding fpring water place fide. William or Williham, the fpring water place vale or home. Wilts, the fide fpring place, wilds or woods ; woods being expreffed as the fprings of vegetation or growth. Wilton or Elandon, Wilts, Norf Heref. a fpring water place fide or bank town. Wimbledon, Surry, the furrounding fpring water place. Wimondham, Norf. on the furrounding fpring water fide vale. Wimondley, Herts, on the furrounding fpring water fide place. Winna or Winn, the inclofing fpring. Winnander Mere, Lanc. the inclofing fpring water part lake or mere. Winceby, Linc. the inclofing fpring fide part or habitation. Winchelfea, Suff. the inclofing fpring at the fea fide. Winchefter, Winton or Vindolana, Hants, N-hum. Scot. the inclofing fpring city, town or bank fide. Winchingdon, Bucks, the inclofing fpring water confines town. Winco Bank, Yorkf. the inclofing fpring water confines bank. Windfor or Windlefaire, Berks, on the inclofing fpring water place fide. Windrufh, Oxon, the lower inclofing fpring water part. Windugledy, Vindogladia or Winbourn, Dorfet, the inclofing fpring water place fide country or neighbourhood. Wincaunton, a town on the inclofing water confines. Winfted, Yorkf. the inclofing fpring water fide or ftation. Winfter, W-mor. the inclofing fpring fide or lower part or country. Winterton, Norf. a town on the inclofing fpring fide or a winter town. Winwick, Lanc. the inclofing fpring water confines or ftreets. Winwidfield, Yorkf. the inclofing fpring fide or edge vale or dwelling place. Wifredsfleet, the fide fpring part water place. Wirhall, Chef. the high fpring water place or hall. Wirkington, Cumb. the inclofing fpring water confines town. Wifbich, Camb. the fide or lower fpring water part confines. Wifk, Yorkf. Wales, Scot. Ire. the fide or lower fpring water. Witenham, Bucks, a vale on the fpring fide. Wintering, Suff. the fpring at the inclofing water part. Witham, Linc. the fpring water fide vale. Witherenfey, Yorkf. the fpring water on the fea fide ; Witlord and Witley, Worc. Wales, the fpring water bank fide, land or place. Wittlefmere, Hunt. at the fide fpring water place mere. Wivelcomb, Som. the fpring water place vale, habitation or commot. Wiverby and Wiverton,

4

Wiverton, Nott. the spring water part abode or town. Hence
Wick, the inclosing spring water confines; Wicket, Wicker
and Wicked, on or at the inclosing spring water confines;
Wide, the side spring; Widow, the side spring, wife or the
spring dwelling or being part; Wig, the spring water confines
or covering; Wight, at the inclosing spring water confines;
Wild, the spring water place side; Wild, an illation or emanation of the spring of light; Willow, on the spring water place;
Win and Wine the inclosing spring; Wince and Winch, in
the inclosing spring water; Wind, the spring in the sides or
air or the surrounding spring; Wing, on the spring; Wink,
a spring or action on the side; Winter, the spring in part;
Wipe, the spring part; Wire, the animal spring water or
milk; Wise, the seeing spring; Wish, the lower side or female spring; Wisp, the spring side part; Wist, the spring
lower side; Wit and With, at the spring or the side spring,
meres or boundaries, which ascertain petitions; Witch, the
side spring water confines; Withe, the side spring or spring
side; Within, in the side spring; Without, out of the side
spring; Witness, the side spring in the lower part or border; Wizard, the seeing spring art or on the side spring water
side.

Woburn and Wobury, Bedf. Heref. on the circle or surrounding spring water or dwelling part or burrow. Wockey
Hole, Som. the inclined spring water hole. Woden or Boden,
Wilts, on the surrounding spring or water side part vale, man
or warrior. Wodensdike, the ditch on the surrounding spring
side. Woken, Setna, on the surrounding spring confines setting division or district. Woking, Surry, the surrounding
spring water confines. Woldsbury, Wilts, the surrounding
spring water place side or wilds dwelling place. Wollaten,
Nott. a town on the surrounding spring water place. Welo-
ver, N-hum. the surrounding spring water place. Wolvey
and Wolver-hampton, Oxon, Staff. the surrounding spring
water place dwelling or vale part town. Womer, Herts, the
surrounding spring water or river. Wondy, Wales. on the
inclosing spring side or coast of the Wenta. Woodbridge,
Suff. the surrounding spring water part side or bridge. Wood-
chester, Gloc. the surrounding spring side city, castle or fortification. Woodcot, Surry, the surrounding spring side
coast. Woodford, Dorset, the surrounding spring side way or
ford. Woodstock, Oxon, the surrounding spring side station.
Woolwich, Kent, the surrounding spring water place street or
fortification. Wootton Gate, Oxon, on the surrounding spring
side gate. Worcester or Wigornia, a city on the surrounding
spring water or the spring water border. Workfwoth, Der.

by the furrounding fpring water fide. Wormegay or Wrongey, Norf. the furrounding fpring water confines. Wormleigh, War. the furrounding fpring water confines or worm hunting place. Worfley, the furrounding fpring water lower fide or ifland place. Worfted, Norf. the furrounding water ftation and worfted. Worth, Surry, Gloc. Staff. Lanc. the furrounding fpring water fide or habitation or by the fide. Wothorp, N-ham. the furrounding fpring gate. Worthington, Lanc. a town on the furrounding fpring water fide or by the border water fide. Wotton, Surry, Staff. the furrounding fpring fide town. Wye, Wales, the fpring or river. Hence Wo, from fpringing or the fpring from; Woman, the fpring part or from man; Womb, the furrounding fpring part of man; Wonder, the furrounding fpring in things; Woo and Wore, on the furrounding fpring of animals; Wood, the furrounding or fide fprings; Wool, the fprings on an animal or fheep; Word, the divifion fpring of man; Work, on the furrounding fpring water or border confines; World, man's fide place or orb; Worm, the furrounding animal fprings; Worfe, below or at the furrounding fpring fide; Worth and Wort, the furrounding fpring or human fides or parts; Wound, dividing the furrounding fpring water fides.

WRAYSHOLM Tower, Lanc. the fpring water part fide promontory or ifland tower. Wragby, Linc. the inclofing fpring water part or habitation. Wrech, Wreke, Wreken, Wrenoc and Wre, N-hum. Salop, Wales, on the inclofing fpring water confines or brookes. Writtle, Eff. the fide fpring water place fide. Wrothes, Wrotham and Wrottefley, Kent, the furrounding fpring water fide, vale, place or dwelling place. Wroxeter or Uriconium, Salop, a caftle or city, on the furrounding fpring water part. Wye, Kent, Surry, Gloc. Der. Heref. Wales, the fpring water or a river. Wyr and Wyrwy, Lanc. Wales, the fpring water or water fpring. Wyth, the furrounding fpring of wind or air or the fide fpring of water. Withcomb, Devon, the fide fpring or fpring fide vale or dwelling. Wythgrig or Mold, N-Wales, the fide or furrounding fpring upper confines water place fide or hill. Wythva or Snowden, N-Wales, the furrounding fides or hills or fnowy parts or dens. Hence Wrack, the fpring water confines. Wrangle, on the inclofing fpring water place; Wrap, the fpring water part; Wrath and Wroth, at the fide fpring water part; Wreak, the inclofing fpring water part confines or wrath action; Wreath and Wrythe, a thing on the fpring water fide; Wreft and Wrift, the leffer fide fpring part; Wreftle, on the lower fide fpring water place; Wretch and

Wrighed

Wright, the fpring of man at the inclofing water or border ; Wring and Wrong, from the inclofing fpring water confines; Writ and Write, at the fpring water fide ; Wry, the fpring.

YALE, N-Wales, above the fpring water place or vale. Yanefbury Caftle, Wilts, the inclofing fpring water part or burrow caftle. Yardley, N-ham. the fide fpring water place. Yare, Norf. the fpring water or river. Yarmouth, the Yare mouth or the fpring water at the fea fide or a haven. Yarum, Yorkf. the furrounding fpring water. Yda, the part or fpring water part. Ydron, Ire. the inclofing fpring water. Yverin or Adgebrin, N-hum. at the inclofing fpring water part. Y-kenild Street, War. Staff. on the inclofing water fide or Kentifh ftreet, Y-kill, on the inclofing water place or the water place fide. Y-linguin, N-Wales, the inclofed fpring water place lake. Ynis Enli, N-Wales, in the fide water place or an ifland in the flood. Yoghal, Ire. the high furrounding water place or fea. York or Eboracum, the circle water or border confines or city. Ytradelyde, the fpring water place fide ftreet. Yftwith, S-Wales, the fide fpring lower fide. Yrwith, the fide fpring. Hence Yacht and Yatch, the fpring water fide veffel ; Yard, the fide fpring water ; Yarn, inclofing ; Yar, on the fpring ; Yawl, on the fpring water place ; Yawn, on the fpring ; Ye, the fprings of man ; Yea, the fprings of earth and water ; Yean, in the fpring ; Year, the fpring on ; Yeaft, the fpring on the lower fide or furface ; Yelk, the fpring water place inclofed ; Yellow, from the fpring of light ; Yelp, the call fpring part ; Yeoman, the furrounding fprings man ; Yerk or Jerk, on the inclofing fpring water confines or action ; Yes, the feen or fide fprings ; Yefterday, the fpring at the fide or part divifion or day ; Yet, at the fpring ; Yield, the fpring water place fide ; Yoke, the circle fpring confines ; You, the furrounding fprings of man ; Yore, the fpring from or off ; Young and Youth, in the fpring ; Your, man's furrounding fpring.

☞ It may not perhaps be improper here to obferve, that the circle of primary names will, by a deliberate attention to the various fides and meanings of confonants and the different ways of their being marked by vowels, appear to be of fo large an extent, as exactly to correfpond with the circle of things : as for inftance ợ the general fign or name of a fpring or what fprings when put to r, thr, expreffes the fpring or pre-eminent-ly man and thence worth, virtue, truth, &c. as the qualities of man ,

man; if it be put to *er*, *the water*, it will denote the spring of water, seasons, &c. and their qualities ; *v* on *ale* and *alley* will signify a place on the spring water ; on *ille*, it expresses the flowing spring water or dwelling place ; its auxilliary *w* with *ill*, the place of light, means the spring of light or the human will ; Womb, is the spring part of woman ; world, man's side or surrounding circle or orb, and various others, that have been already explained. Then the different sides or prospects, in which the mind views those primary names, greatly multiply their signification, as do the various ways of compounding them, and so as to the other vowels. Besides, nature seems to be but combinations of few analogical systems, made up of units, dualities, and triads, with the quaternion of things or sides variously disposed. We have here closely adhered to the primary, hieroglyhpic, and argraphic, meaning of names, though the more removed sense hath been occasionally added, but the former definitions having been promiscuously derived from the different sides, prospects, and meaning, however they may discover the meaning of words, must be somewhat less perspicuous than those of the present essay.

AN UNIVERSAL

ENGLISH GRAMMAR.

GRAMMAR of a particular language is the art of rightly speaking and writing it according to its own established and fixed rules. An Universal Grammar is the form and mode of rightly expressing our ideas by such signs and sounds as correspond with ideas and things, as their natural archetypes.

NAMES consist of certain lines or chords, springs, circles, semicircles, surfaces, sides, divisions or parts, figures and letters, which as symbols and articulate sounds or voices, describe and represent the elements and parts of nature; their combinations into particles or syllables of two letters; models of names of two particles or three letters, with an edition of the other; and decomposites or still larger combinations of these in names, phrases, and sentences.

OF LETTERS.

The letters in general use are the following: A, a; B, b, C, c; D, d, dh; E, e, F, f; G, g; H, h; I, i, j; K, k, L, l, ll; M, m; N, n, m; O, o; P, p, ph; Q, q; R, r, rh; S, s, sh; T, t, th; U, u, u; W, w; X, x; Y, y; Z, z; whose powers, founds, and meanings, are illustrated under the proper heads, as well as in the former essays on grammar. But these letters seem to be more properly distinguishable according to the following sorts and order, namely O, a, e, i, u, w, y, the vowels or elements of language; p, b, f, ph, v, m; c, ch, g, k, k, q, x, wh; rg, t, th, d, dd, n, the mutable, subsidiary and auxiliary consonants; and h, l, ll, r, s, th, z, compounds of the two former sorts. And this distinction of vowels and consonants arises from the formers,

* B   being

being compleat articulate founds or voices, and reprefentations of elements and general objects; and the later, though in fome degree they include the figure and founds of vowels, even in conjunction therewith, become the voices or reprefentations only of the parts of elements or vowels

THE confonants p, c, t, are radical and mutable, inflectory or fpringing, and b, f, p, ph, v, and m are the fprings or inflectories of p; but ph, and v, are fo only as auxiliary to f. The ch, g, j, k, ng, q, wh, x, and fometimes fh are the inflectories of c; but the j, k, q, wh, x, fh, are fo only as auxiliaries of g and ch. The th, d, dh, n are the inflectories of t; but b, d, g, m, though not radically initial mutables, are fo in the fecond and third degrees. And ll and rh are rather afpirates of l and r than inflecting letters; and thofe variations of the confonants may perhaps be fomewhat illuftrated by the following Welfh form thereof of Dr. Davies.

1. Mutable radicals.	2. Soft.	3. Liquids.	4. Afpirates.
1 Car	Gar	Ngar	Char
Pen	Ben	Mhen	Phen
Tad	Dad	Nhad .	Thad
2 Bara	Fara	Mara	
Duw	Dduw	Nuw	
Gwr	Wr	Ngwr	
3 Llaw	Law		
Mam	Fam		
Rhad	Rad		

Or the following fpecimen of the antient variations or inflections of the English, which are become obfolete

Mutables.	Soft Springs.	Liquids or fluids.	Afpirates.
Pale, pall	Bale	Male	Fale, Vale
Par	Bar	Mar	Far
Park	Bark	Mark	Fark, Far-ce
Pell	Bell	Mell	Fell
Per	Bear	Mer	Fear
Pit	Bit	Mid, meet	Fit
Pull	Bull	Mull, Mule	Full
Pan	Ban	Man	Fen
Pafs	Bafs	Mufe	Fafs
Tear	Dear	Near	Thare
Ten	Den	Nen	Then
The	Die	Nigh	Thigh

Tile	Dil-ate	Nile	Thill
Tir, Tri	Drey	Nigher	Three
Time	Dim	Nim-ble	Thyme
Tin	Din	Nine	Thin
Tomb, Tome	Dome	Nome	Thong, Thome
Two, To	Dwi, do, Duo	No-ble	Tho
Cam-p	Game	Name	Cham-pain
Can	leGan	Han	Chan, Shan
Cap, Cape	Gap, Gape	Knap, Nape	Chap, Chape
Car	Gar-den	Nar-row	Char, Char-m
Cafe	Gafe	Nafe, Nafe	Chafe
Cat, Cad	Gad, Gate	Nat-ive	Chat
Cell, Kell	Gelly	Neal or Nel	Shell
Comb, Cwm	Gum	Num-ber, Home	Chum
Cor-d	Gore	Nore	Chore

And thefe variations or fpringing of the confonants originally ferved to diftinguifh fubftances, qualities and actions, genders, perfons, and cafes both as to fenfe and founds, and furnifhed mankind with the means of giving names to ideas and things, as they fhould be enlarged on the firft principles of fpeech, as well as fuch as more corruptly fprung therefrom like Humber and Umbria, for Chumber and Cumbria, vixit for bixfit, and Militavit for Militabit.

Though letters have been already explained in a former effay, we fhall here add the following illuftration thereof, and firft as to vowels. The o then as a circle is expreffive of a globe, the world, an orb, a continent, an ifland or other leffer circles or parts furrounded by water or otherwife, dividing the boundaries and borders of countries, the buttock joints or articles of the human pair or microcofm, the earth and water parts of man joined like the hemifpheres in a circle, and other circular fhuts or inclofures both of matter and motion and fpace or extenfion; and in fome dialects the propofition _from_, as the fide or limit is the extremity of and greateft diftance from any part or country. The a and e fignify the hemifphere of this globe of earth and water, the elements of earth and water, the earth and water parts of man or the parts that diftinguifh male and female, as vowels, the definitive articles a and e and the mafculine and feminine genders; the e being by itfelf a privative, t or its inflectory th fignifying a fide or the fide of man and _the_, the feminine gender. The letter i is the element of fire and all its qualities, which by its rays, dots and lines penetrates bodies, and caufes the flow and fpring of things and alfo light, and it fignifies man, as a

lighted

lighted candle, and an upright, springing or living line and also length, the pronoun i and the first person singular or man. The letter u is the return of its spring in the seasons, water, vegetation, generation, nutrition, animals, man, truth, knowledge, virtue, voice and other things, and also the pronoun wo, the first person plural or man and woman ; w and y are auxiliary expressions for the spring of springs and vegetation, and the latter in some dialects stands for the articles a and the and represents man on his head, with his legs extended after his fall in Eden, and the spring division of countries, and other springs about which he was to labour and sweat for the loss of spontaneous growth The a, i, u, are the masculine gender, the e is feminine and the other vowels are neuter.

The radical inflecting c or the half of o, with its inflectories or auxiliaries ch, g, ng, h, k, q, sh, and x or cç signify the water, sea, female and other surrounding sides, shuts, inclosures and circular things, with their actions, qualities, and motion. The letter p is expressive of material or inanimate parts, b of living, animal or vegitable parts and f, ph or v of their qualities, springs, motions or inflections ; as for instance, p is expressive of certain parts or pieces of earth and water called a river and the virile member or a penis, b signifies the same in their spring, and f and v are expressive of their fluxion and motion to the marine and female reservoir, and the letter m signifies surrounding and bulky parts, as a man's body, a mountain, the sea, as surrounding a country and the circle of things and motions about us. The n is an inclosing part or side or a part or side inclosed and as such a negative or unseen. The t hieroglyfically represents man's legs and thighs close together with the toes outwards, the water place of man and a river flowing into the sea from u, the earth ; p being the thigh article or part, l, place, or part of space, and n its other inflectory, [the part between the thighs] when extended or the inclosed part of a country The letter t is also expressive of extension of parts, sides, surfaces, coverings, possessions and things ; as its inflectories th, d, dh and n of their various parts, divisions and qualities, and also the article the and affirmations by the surrounding sides or things, as is s by the senses and things seen and founded and d of the division of things and qualities visible and invisible and privatives as spirits, the division or creation of the world and the separation of earth and water. The r signifies various parts of earth and water. The l signifies place and its sorts and degrees, as altitude, the different surrounding sides, a river or other water places, as that between a persons legs, the water place of man. And the capital letters and numbers, represent the several divisions of countries or possessions; as A a compound of A and V, the first division from the sea of the river; M and W, the same repeated on the sea side to a thousand; B and F, the same on the boundary river sides; C and G, a canton or hundred or side of a circle; O, a circle; D, a single division on the side; E, the river divisions falling into the sea; I, a flowing river; T, the same with its two sides falling into the sea; L, one side thereof; and Y, a division of a river or opening like the Nile at the sea or a tree or other vegetable growth. Hence letters appear to derive their origin from double prototypes, namely, the world of earth and water and their various parts, as the river and sea; and the mikrocosm of man and woman with their parts, as the penis and vulva, and tongue, as to voice, &c. And as to the sounds of consonants, they are better expressed with their vowels in particles, than by any other practicable method of doing it.

Of PARTICLES and SYLLABLES.

Particles or syllables are compleat signs and articulate sounds voices or names, either of single vowels or compounds of a vowel and consonant, which in their meaning exactly correspond with that of the consonants, as they have been already explained, so that the joining of vowels to consonants, in order to form particles, serves no other end than to mark or distinguish to which of the elements or vowels the consonants are applied, without adding any thing to their powers or value either in sense or sound. And as the joining of vowels to consonants, for the formation of particles, is attended with very considerable unnecessary expence, trouble, prolixity, doubts and uncertainties, an attempt to reduce the three sorts or distinctions of vowels, consonants and syllables, to two sorts or classes of vowels and consonants articulate, may not perhaps be deemed altogether as useless to the design of an universal language.

Then points and commas, or the half of them, as symbols of the sun and moon rising and setting, might be fixed at the lower, middle and upper parts, on the sides of consonants projecting to the right and left occasionally, to mark the active and passive senses, in the same manner as vowels appear to do by their being placed before or after the consonants. As, e, or v active, for instance, might be supplied by a comma placed at the lower, middle or upper parts of the left side of consonants, and u, o, or inactive, in like manner by points; and to denote a passive or substantive sense, they might severally be fixed on the right sides of consonants in the like manner, and both may alfo be marked with a point for the w. Then the vowel claffes, in lexicons, will confift of actives and the reft of fubftantives. And the following explications of particles or fyllables muft be a fufficient guide therein.

Ab, pa; Ep, pe are the dead, cold and inanimate parts or divifions of earth and water and privatives. Ab, ba; eb, be are the living, animate, fpringing and vegetable parts of earth and water, and male and female, as a fon, a lamb, a buck, a bean, and their privatives. Af, fa; ef, fe exprefs the qualities and effects thereof, as thofe of a father, a man's faith, an animals fear or ferocity, a plants fertility, and their effects and privatives Am, ma; em, me are expreffive of the furrounding parts or exiftences of earth and water, feas, rivers, mountains, hills, man's body and garment, inclofures and boundaries of countries, and things promifcuous and their privatives.

Ac, ca; ec, ce are earth and water, male and female fhuts, inclofures or borders, and their actions or modes of local motions. Ag, ga; eg, ge are the effects and qualities thereof in vegetation, growth and generation, as eggs, egg, egiſment, egin or fhoots of corn or young animals, egret, and their efforts to motion and life. Ang, nga; eng, nge are the inclofing, ftreightened or fhut water parts of a country or other places, as angle, anger, anguifh, enclofe, engage, engine, eng or extenfive and England. Ach, cha; ech, che are an higher degree and forts of actions and fhuts and their returns and effects, as ache, achieve, chafle, chat, chaw, echo, check and their privatives.

At, ta; et, te; ath, tha; eth, the, a combination and extenfion of parts, fides, furfaces, countries, coafts, boundaries and poffeffions, the prepofition *at* and the article *the* or the fide, with the things and qualities therein, and affirmation by the furrounding fides or things. Ad, da; ed, de; add, dda; edd, dhe the divifion of earth and water, and qualities both vifible and invifible, privatives, and the verb *add*, or an addition, only by dividing but not in quantity. An, na; en, ne are parts of earth and water, man, or things in exiftence or poffeffion, and feen or fhut out of poffeffion, and unfeen and a negative of exiftence.

Ip, pi; ib, bi; if, fi; im, mi fire, high, living, and generative parts, and the power or his heat and qualities on the furrounding parts and things in general.

Ic, ci; ig, gi; ich, chi the firft, chief or upper motions or actions of heaven light, and their actions in vegetation and generation, the upper covering, or fky and other inclofures and fhuts, and the prepofition on.

Ii,

It, ti ; ith, thi the properties and qualities of fides or things from fire, the fide man or thing and the pronoun *it*.

Id, di ; idd, ddi fignify the divifions of fire, as light, fight, colours, difcernment, the human intellects, fpiritual qualities and privatives thereof.

Op, po ; ob, bo ; of, fo ; om, mo the circle parts of the earth, water, man, animals, and country borders and the prepofition *of*.

Oc, co ; og, go ; ong or on, ngo ; och, cho the circles of water, fire and motion, the fea and other water boundaries, fhuts and inclofures, the prepofition *on* and the verb *go*.

Ot, to ; oth, tho', the furrounding fides, circles, borders, and coverings, the prepofition *to* and the adverb *tho'*.

Od, do ; odd, ddo divifions of the circles of motion or actions both prefent paft and future, and the verb *do*.

Up, pu ; ub, bu ; uf, fu ; um, mu the various furrounding fprings of things upwards, as thofe of waters, vegetables and animals, and other parts and living things and animals as to generation and *up* and *upon*. U'c, cu ; ug, gu ; ung or un, ngu ; uch, chu the chief fpring of animals in generation, and other chief fprings and their negatives and *unto*. And the particles formed of k, q, x, wh and fh are auxiliary to thofe formed of the c, and its other inflectories.

Ut, tu ; uth, thu, the fide fprings, a woman and the pronoun *thou*.

Ud, du ; udd, ddu, the divifion of fprings, or the returns of earth and water and man in feafons, vegetation, animation, generation and intelligence, and their negatives and privatives.

Al, la ; el, le ; il, li ; ol, lo ; ul, lu, the place of earth, water, fire and light, the circle of matter and motion and man, on a place, below a place, a high place, all place, and the place of the human light or intelligence and *all*.

Ar, ra ; er, re ; ir, ri ; or, ro ; ur, ru, the earth, water, fire, world, fpring and their fimilar parts, properties, accidents and qualities and negatives, as upon the flowing or ftanding rill of water and fire and motion, a water border, from, and the various fprings of animals and things.

As, fa ; es, fe ; is, fi ; os, fo ; us, fu, the earth, water, fire, a circle or border and fide fprings or things furrounding, feen, fmelt or founding, low and lower and male and female, affirmations by the fenfes of things feen and furrounding, far or at the fide or diftance, the human fpring, a conjunction *as*, the affirmation *is*, the adverb *fo*, and the pronoun *as* or *us*.

As,

Ao, oa ; eo oe ; io, oi the circle and motion of earth and water, the fun and all things, and their shuts or privatives.

Ai, ia or jah, the cause of fire and its motion or emanation on the earth. Y, a spring thereof from the earth.

Ah, ha; eh, he ; ih, hi ; oh, ho ; uh, hu are the different forts and degrees of energies, emotions and paſſions, correſponding with the meaning of the vowels. And all the foregoing definitions of names correſpond with, and ſupport theſe definitions of letters and particles.

Of the MODELS of PRIMITIVE NAMES.

The firſt models of compound names were combinations of two ſyllables or a vowel with two conſonants and an eliſion of the other vowel, the remaining vowel being like a, e, o and i, u, y, ſimilar and ſufficient to diſtinguiſh the element or genus, and to direct the application of both conſonants ; and the larger forts of names are nothing more than decompoſites of thoſe models, as appears by their explication in the foregoing work.

Thus far perhaps the frame and conſtruction, or grammar of the primitive language extended ; or at leaſt the letters and ſyllables of all languages, according to the ſenſe here given thereof, have always been, and are ſtill univerſally the ſame. And were the models of Engliſh names properly examined and tried on the Engliſh form of inflection, they might appear to be primitive and copious, both as to radicals and inflectories, and ſuch as have been loſt might thereby be diſcovered, and found out in other dialects ; always obſerving in their choice that their frame and conſtruction agree with our explication of letters, particles and models, and leaving out the terminations, which in moſt languages are arbitrary. But if after all our own language ſhould appear both as to its parts and conſtruction to be the neareſt of all dialects to the primitive language, and the fitteſt to be adopted for an univerſal one, from the preciſion and copiouſneſs of our radicals of places, of which thoſe of things are genuine copies, the preciſion of ideas, diſpoſition of perſons and things, as well as the climate correſponding, it may perhaps be deemed unneceſſary to attempt the retrieval of languages from their preſent advanced and complex ſtate and progreſs with the arts and ſciences to their primitive form. Hence, other forts of grammatical names, or parts of ſpeech, are become requiſite for the conſtruction of a rational univerſal language, ſo as to expreſs ideas, paſſions and things rightly and adequately in their preſent enlarged ſtate ; and being unable to furniſh ourſelves with

any

any other names fitter for the purpoſe: we ſhall here make uſe of the following Engliſh grammatical names or parts of ſpeech, namely, the articles, nouns ſubſtantive, adjective and participle, pronouns, verbs ſubſtantive, active and paſſive, adverbs, conjunctions, prepoſitions and interjections.

Of the ARTICLES.

The Engliſh articles are *a*, *an*, or *any* and *the*, the former of which ſignify the earth, man, maſculine gender and ſingular number, and *the* is expreſſive of the ſide water, woman, the feminine gender and plural number in the order of the creation ; and being as ſuch of a ſubſtantive nature, they, when prefixed to other names whether of ſubſtances or qualities by their coaleſcence therewith ſubſtantive, or denote them to be ſubſtantives and expreſs their genders and numbers ; as, *a* man, *an* horſe ; *the* woman, men ; *any* man, men, woman, or thing ; but not *a* men, *an* horſes, or *the* one, as woman implies a plural. Hence we diſcover by their meaning, the firſt ſort to be of the maſculine gender and ſingular number, the ſecond to be the feminine gender and plural number, and the third to be of the neuter, or neither gender.

Of NOUNS SUBSTANTIVE, ADJECTIVE, and PARTICIPLE.

Nouns ſubſtantive are the names of ſubſtances, or ſuch things as we conceive completely to ſubſiſt either by themſelves or in concord with the articles, without the aſſiſtance of any other names ; as man, a play, an act, any thing, the ſpirit.

Nouns adjective and participle on the contrary are the names of actions, qualities and attributes, added to ſubſtantives to expreſs their qualities and accidents, and without which, being in their nature indefinite, they cannot ſubſiſt ; but the participle is in ſome caſes diſtinguiſhable from the adjective as having the particles *ing* and *ed* added thereto, to expreſs the preſent and paſt time ; as, a wiſe man, a wicked play, a virtuous act, the holy ſpirit, a loving or loved thing, or I am loving or have loved.

Nouns ſubſtantive have the diſtinction of the maſculine, feminine, and neuter genders, and the ſingular and plural numbers, by their own meaning, the coaleſcence of their articles and the addition of *s* or *es* to the ſingular or a variation of the radical vowels a and o into e to term the plural number ; as for inſtance, a man-men or women, foot-feet, bull-bulls, cow-cows, city-cities, determine the ſingular and plural numbers and maſculine and feminine genders ; but people, cattle, calves, fiſhes, corn, graſs, and other general names

be-

being originally fo framed, exprefs the plural numbers and neuter or neither of the former genders; and had not the female
occafioned the diftinction of fex, fubftantives would have had
no diftinction of genders; whatever, therefore do not appear by
their meaning to be of the feminine or mafculine genders
ought ftill to be deemed of no diftinguifhable or neuter genders. But as to the diftinctions of common and proper names
or the declenfions of cafes, there are none either in nature or
the Englifh language, except that propofitions in compofition
with verbs, or appofition to fubftantives or as has been already
fhewn in a former effay, their pofition before, and after the
verbs exprefs the nominative and objective cafes.

Nouns adjective agree with the fubftantive in gender and
number without any change or variation. But they exprefs
their different degrees of qualities, diftances, quantities, and
attributes, by the addition of the particle *er* to the pofitive to
form the comparative, and *eft* the fuperlative degree; as, far-
farther-fartheft; great-greater-greateft; high-higher-higheft;
much or many-more-moft or mo-eft; fwift-fwifter-fwifteft;
fweet-fweeter-fweeteft; or where the adjective has been compounded beyond its primitive form, by prefixing thereto fome
of the adverbial names of comparifon, as many, more, or moft.

Of PRONOUNS.

Pronouns are general nouns, which in concord with verbs
perfonate, and exprefs perfons and things with their number
genders and relations; and thofe of the Englifh language are
as follows:

	Perfonals.	Poffeffives.	Relatives.
Singular	1. I, me, myfelf;	my own, mine;	one, any, none.
	2. thou or you, thee, thy, thine; thy felf;		this, each, every, ei-ther, thefe.
	3. he, fhe, it, him, his, hers, its; her, himfelf, her-felf;		that, fome, another, fuch
Plural	1. we, us, ourfelves; our, ours;		who, whofe, whom.
	2. ye or you, your-felves; your, yours;		which.
	3. they, them, them-felves; their, theirs;		what, thofe, others.

Thefe little fignificant names are all feveral originals, and
like nouns, exprefs the cafes in pofition with verbs without
any

any variation, and their genders as they perſonate man and woman, and their poſſeſſions ſubſtantively are of the maſculine
and feminine genders, and the relatives and adjectives being
general names and expreſſive of neither the one or the other,
but of both alike, are of the neuter or no gender, and ſuch of
the poſſeſſives and relatives as ſtand by themſelves, and will
not like adjectives join and coaleſce in concord with ſubſtantives,
are themſelves ſubſtantives, and the reſt are adjectives; and who,
which, what, whoſe, whom, are called interrogatives, though
other names ſeem to be equally applicable in that ſenſe, and
the relatives expreſs perſons, things, actions, or parts of diſcourſes or ſentences. But the following explication of theſe
names may perhaps illuſtrate their uſes and properties in language better than any rules that can be laid down for the purpoſe.

1. *I* perſonates man as an upright ſpringing or living line or
a lighted candle-*me mi* or *my* repreſents him with a body and
ſurrounded by matter-*myſelf* expreſſes him as within himſelf or
in his own ſide-*my own, mine*, within his circle ſpring-*one, any*,
the circle or incloſing ſpring-*none*, its negative.

2. *Thou, you*, perſonate the ſide, divided or ſecond ſpring or
woman--*thee*, the ſide i or woman-*thyſelf*, woman in herſelf or
her own circle-*thy, thine*, a woman's incloſures or properties-
this, theſe, the female ſide or ſides or incloſures-*every, either*,
the ſide or incloſing ſpring.

3. *He* or *hi*, the upper acting man or the male or flowing
ſpring-*ſhe*, the lower or female of the ſex-*it*, man or woman's
ſides, things or poſitions-*him*, the higher and moſt powerful
acting man *her*, the weaker or more paſſive *himſelf*, he in his
own circle-*herſelf*, ſhe in hers-*his*, the he or male ſide-*hers*,
her ſide-*that*, the ſide things-*ſame*, the ſurrounding or circle
ſides-*its*, a ſide thing or an exiſtence-*another*, one below the
circle ſide-*ſuch*, the upper ſide or the ſky.

4. *We*, the two ſprings of man and woman-*us*, the ſide,
divided or double ſprings of man and wife-*ourſelves*, man and
wife within themſelves-*our*, the circle of man-*ours*, man's ſide
circle-*who* the upper ſpring or what man? *whoſe*, the man's or
what man's ſide-*whom*, what, or the ſurrounding ſpring or
man.

5. *Ye, you*, the divided or from the ſide ſprings or women-
ye rather, their circles-*your, yours*, their ſide ſprings a ſuch, the
above acting ſpring.

6. *They*, the ſide ſprings or mankind-*them*, the ſurrounding
ſides-*themſelves*, the ſide ſurrounding circles-*their, theirs*, the
ſides

ñdes *what*, the ſide ſpring or thing-*thoſe*, the circle or ſurrounding ſides or things *others*, below the circle or border ſides.

Of VERBS.

Verbs are of three ſorts, namely, ſubſtantives, actives, and
paſſives.

The ſubſtantives are bare affirmations of exiſtences as, I *am*,
be, or *have been*.

The actives affirm with an action, quality, or attribute of
a ſubject; as *walks*, man is in action, *I touch thee*, thou *feel-
eſt* me, or a man *is active* and ſenſible.

The paſſives are the reciprocal aſſertions of the latter by the
ſubſtantive verb and an adjective or participle preſent, paſt or
future; as, I am or have touched or ſhall touch thee, thou
doeſt or ſhalt feel or haſt felt me, I am feeling or I received the
touch. And the verb ſubſtantive is an auxiliary in the conjugation of other verbs.

Verbs as affirmations of exiſtencies, actions and paſſions
with their times, diviſions or ſucceſſions in place coaleſce in
concord and government with the perſons, numbers and genders of nouns and pronouns or ſubjects in the following manner, mode or form.

The MODE of Conjugating VERBS.

Numbers	Persons		The Present.		Past.		Future tenses or divisions.	
			Absolute.	Conditional.	Absolute.	Conditional.	Absolute.	Conditional.
Singular.	1	I	am, be, have, do, love, teach, read, heard, go	may, can would, could, should or ought to have, do, love, teach, read, &c.	was, have been, had, did love, taught, heard, went	might, could, would, should, or ought to have been, had done loved, taught, heard, gone	shall, will, or must be, had, done, loved, taught, heard, gone	shall, will, or must have been, had, done, loved, taught, heard, gone
	2	you or thou	are or art be or beest have or haft do or doft love or lovest, &c.	- - - - -				
	3	he	is, hath, doth, loveth or loves, &c.	- - - - -	was or hath been, had, done, loved, taught, read, heard, went			
Plural.	1	we	are, be, have, do, love &c.	- - - - -	were or had been, had, done, loved, taught, read, heard, went			
	2	ye	- - - - -	- - - - -				
	3	they						

Imperative mool.

Singular.

1 Let me be, have, do, love, teach, &c.

2 Be, have, do, love, teach thou, &c.

3 Let him be, have, do, love, &c.

Plural.

1. Let us be, have do, love, &c.

2. Be, do, have, love ye, &c.

3. Let them be, have, &c.

Infinitive mool.

To be, have, do, love, &c.

To have been, had, done, loved, &c.

Participle present.

Being, having, &c.

Past.

Being or having had, loved, &c.

These auxiliary verbs are divers, invariable, energic affirmations of present, past and future divisions or successions of motions or actions in place called times, and the minuter divisions thereof are more precisely distinguished by those additional auxiliaries called adverbs. These are not the modes as some have imagined but the parts that affirm the subject and predicate in prepositions, and modes are the manner of putting the several parts together ; as, he is a wise man that adores and explores his creator and is loving to and loved by his friends and neighbours. And thus by joining the participle and past active to the auxiliary verb *is* are formed the passive and neuter verbs.

As verbs active lose their qualities or action in the past tense, most of the English verbs express the tense, by *ed*, *d* or *t*, as they denote a privative and a contraction of the sense and found from those of actions or qualities to substances, and these that are thus regular in their conjugation are deemed to be regular verbs.

But the following definitions of the auxiliary verbs seems to give a more clear and certain illustration of the same and their real use in language.

Am, the surrounding parts and I *am*, is man springing, living, or existing therein, and affirming the same by his own senses---b being the member that distinguishes the virile sex, *be* are those of both sexes and *I be* is man's springing, feeling, and affirming by the same---*May*, the surrounding springs and motions and by *I may*, man asserts a possibility of his or things springing or coming into spring, motion, or action---*Can*, acting in or into the inclosing water place, or possession, and by *I can* man asserts his own ability therein---*shall*, the lower side up, in spring or present and by *I shall* man asserts the springing up into action or possession, as a child in the womb or a root in the earth---*will*, the spring of light or intelligence, and by *I will*, man declares that his mind sees and approves of an action---*most*, the lower sides of the surrounding springs, as the boundary springs or roots of plants below the surface of the earth and by *I must* man asserts their springing up---*should*, the lower side springing up had, aside or past and by *I should* must be meant that mans springing up was at the lower side or past ---*would* signifies that the spring of intelligence or will was had---*could*, the inclosing spring water place or possession past ---*might*, the surrounding spring water aside---*ought*, the circle fixing water or border tide or day---*art*, on the spring and the side on the spring---*wast*, *wilt*, the feminine shall and will---*wert*, the same as to be---*is*, the side or that man exists,

&c.

sees, speaks, &c. *was*, *wast*, a man, woman, or a spring past, below or aside---*were*, the human or other springs past---*been*, the spring part in---*have*, a spring or energic motion of the passions---had or ha-ed, the same at the side, aside or past---*hast*, *hadst*, the female springing up past---*hath*, has, the side spring or man springs and acts---*do*, the divisions of the circle of motion or actions---*did*, the same past or aside---*doth*, the side man's action---*done*, the action in or past.

Of Irregular VERBS.

Such verbs as form the past tense by ed, d or t conformable to the foregoing mode of conjugation being deemed regular verbs, those which vary from that rule have been esteemed as irregular verbs, and rules have been laid down for their conjugation in such their supposed irregular state. But instead of laying down rules for the establishment of error or forcing our language from its natural state of precision to the arbitrary modes and forms of languages with which it has but a very small connection, I shall endeavour to rescue her from such violences, and in the investigation of these sort of verbs, shall follow the distinctions of our best English grammarians, from whom we differ only to follow nature and the sense and meaning of things instead of sounds, which must have been too liable to accidental variations to be the foundation of an universal language.

1st CLASS.

1. Some verbs though ending in ed, d or t, are supposed to be irregular, as having the present or past tenses of the participle perfect and passive all alike without any variation; as let, beat, burst, cast, cost, hit, hurt, lift, light, put, quit, read, rent, rid, run, set, shed, shred, that, slit, split, spread, thrust, wet. But it is observable, that these verbs being originally thus formed like other substantive verbs in the imperative mode, and no contractions of leted, beated, &c. they like the auxiliary verbs, seem to be in their nature incapable of any variations, and as they coalesce with the proper subjects in concord, it seems immaterial whether they change the tenses or not, and ought to be deemed regular verbs

2. Others that vary the past time and participle perfect and passive from the present, are supposed to be irregular, by changing the d into t, or shortening the double into single; as lead, led; sweat, swet; meet, met; bleed, bled; breed, bred; feed, fed; speed, sped; bend, bent; lend, lent; send, sent; spend, spent; build, built; geld, gelt; gild, gilt; read, rent, rend,

b d 3

girt ; lofe, loft. But thefe verbs being originally formed of fubftantives, lofe their active qualities and return to their former fubftantive ftate in the paft tenfe, and the contractions and muta ions ferve to harden the found accordingly, and as they coalefce in concord. they muft be regular in their own nature, and ought to be deemed fo in grammar.

3. Others not ending in d or t, are faid to have been formed by contraction ; as have-had for haved ; make-made for maked ; flee-fled for fleed ; fhoe-fhod for fhoed : and that the following forts, befides this contraction alfo change the vowel ; fell, fold ; tell, told ; clothe, clad. As to the paffive verb *have* it feems to change regularly into ha-ed or had, as do make into ma-ed or made and the reft of that fort. Sell and tell became folled and tolled to correfpond with their fubftantives toll, foll or coll, the tranfaction of telling, felling and calling of cattle at fairs. Nor doth clo-the, clo-eth or ed in cla-ed, or the more contracted clad become irregular. And as to ftand, ftood and dare, durft, being directly from the Saxon ftandan, ftod and dyrran, dorfte, unlefs the purer expreffions can be faid to derive their o igin from the more corrupt, it feems more probable that the Saxon words come from the Englifh, which feem to be complete, without the particle *an* either as fubftantives or verbs, as appears by the foregoing definitions, which comprehend and explain all the fifter dialects of the weft and north, if not of all Europe. And was it true that the Englifh derives its origin from the Saxon, it would have had a nearer refemblance to the German.

2d. CLASS.

1. The irregulars of the fecond clafs ending in *ght*, changing the vowel or diphthong into *au* or *ou*; and fuppofed to be taken from the Saxon, are as follows : bring-brought from bringan-brohte ; buy-bought from byegean-bohte ; catch-caught ; fight-fought from feotan fuht ; teach-taught from tæchan-tehte ; think-thought from thencan-thohte ; feek-fought from fecan-fohte ; work-wrought from weorean-worhte; freight-fraught. Our definitions of thefe fhew the Englifh verbs without the *an*, to be complete and regular in the prefent tenfe except bye, which fhould be wrote buygh but pronounced buy, fo as to form like teach, a regular variation in the paft tenfe of bough-ed or t. And the changing the vowel u, or any other, except a into o, *from* or fhut, to form the paft tenfe feems to be perfectly regular, as alfo to change the a into e but not into *au*, and when teach was pronounced tealh, it regularly formed taught or tough-t or ed, and fo as to work-worugh-t

worugh-t or wrought according to the usual contraction in the past tense and the rest. But the Saxon biot, bot, feot, fut, tæt, thot, sot, wort, are very inadequate to the meaning intended to be expressed. Nor do the asperate h without a vowel or the final e add any thing thereto, and bring may have been formed of bearing the participle of bear, and think may be an original expression for the internal act of the mind, and thought the same past into external actions, as tought the past of teaching. So that with these explanations and corrections, those verbs as they coalesce in concord cannot be properly deemed irregular.

3d. CLASS.

1. The irregulars of the third class are supposed to consist of such as form the past tense by changing the radical vowels and the participle perfect and passive by the addition of the Saxon particle *en*. As to the participle, the best way of correcting its irregularity, seems to be intirely to drop it, as having been at first corruptly introduced into our language by dehgning men, or suffer it to grow into disuse and become obsolete according to the natural bent and tendency of the language to reject all exotics that will not coalesce or agree therewith in concord and to disown them for her offspring, though we must at the same time confess that the Saxon dialects freed from their paint, daubing and disguise will all appear to be her genuine daughters.

2. With respect to the verbs which form the past tense by changing the vowels from a to e, o; e to a, o; i long to the short i, a, o, u; o to i, ew but in no other manner seem to do so according to the sense and primitive mode of conjugating verbs; as for instance, tail, tell; draw, drew; slay, slew, vary from the a positive to the e privative according to the sense; and aawke, awoke, forsake, forsook; shake, shook; take, took, change from the a active to the o shut or past. The variations from e to a and o in get, got; eat, ate; bear, bore; break, broke; cleave, clove; speak, spoke; swear, swore; tear, tore; wear, wore; heave, hove; shear, shore; steal, stole; tread, trod; weave, wove; creep, crope; freeze, froze; seeth, soth or sod; see, saw from the feminine energic e to the masculine a or o short or past, seem to be very natural and expressive. The change of the i long to the short, is expressive of the flowing motion of the long i to be contracted and past in the short; as in bite, bit; chide, chid; slide, slid. Its changes to a and o denote the like motion or action of the long i to be shut or determined in a substance; abide, abode; drive,

drove ; ride, rode ; rife, rofe ; fhrive, fhrove ; fmite, fmote ; ftride, ftrode ; ftrive, ftrove ; thrive, throve ; write, wrote ; bind, bound ; find, found ; grind, ground ; wind, wound ; and its change into a, as in lie, lay ; and into u, as ftrike, ftruck, denote the action of the lower part to have been fprung. The i fhort or beginning of motion changes into u to exprefs its length, fpring or continuance ; as in fpring, fprung ; fling, flung ; ftink, ftunk ; ftring, ftrung ; fwim, fwum ; fwing, fwung ; wring, wrung , ftick, ftuck ; win, wun ; dig, dug ; and into a, as bid, bade ; give, gave ; fit, fat ; fpit, fpat ; as expreffive of motion or action paft. The o into e, as hold, held, fignifies the being deprived of a poffeffion or an inclofed place ; into i, as do, did, or doed, feems to be pretty regular. The long or great o or oo into the fmall or fhort, or rather from w to u, as choofe, chofe : fhoot, fhot, fignifies a fpring with its continuance. And from o to ew, as blow, blew ; crow, crew ; grow, grew ; know, knew ; throw, threw ; flow or fly, flew, fignifies the continuation of a fpringing action or motion. And as thofe verbs naturally coalefce in concord they feem to acquire no other corrections or variations than what have been here intimated and ought to be deemed as regular verbs, as fhould the auxiliaries which are fuppofed to be defective verbs, as divers original expreffions.

Of ADVERBS.

Adverbs are certain auxiliaries added to or fet at the fide of verbs and adjectives to denote the feveral degrees, divifions, and diftinction of times, places, quantities, qualities, motions or actions, refts, affirmations and manner of things in propofitions and fentences ; and which, with their definitions and degrees of comparifon are as follows :

Of place---*here, otherwife, ftraight, upright,* the part a man is upon or in poffeffion of and his fprings, rights and poffeffions ---*above, aloft, atop, up, upwards, longwife,* or longwife, the fpring, the inclofing fpring or fide water or border part or place ---*beneath, below, down, aground, under,* on the circle or fide water part or place---*off, outwards, abroad, around,* at on or towards the circle or border part or fide---*before, facing, onwards, forwards, ahead, throughly,* the circle, border, or confines part or towards or through the fame---*behind, aftern, after,* a man or thing's fide, fhade or hind part---*near, nigh,* on the part upon or on the confines---*far, far away,* on the water part or from the part upon---*amongft, amidft, intermixedly, largely, widely, hither, and thither,* the things furrounding or in the circle of poffeffion or about the country---*there, thereabout,* on or about the

the fide or the water fide---*hence*, on the fide of the part upon
---*thence*, on the fide's fide---*thither*, *thitherto*, *that way*, to the
fide or way there---*where*, *fomewhere*, *any where*, *whither*, *fome-
whether*, *which way*, *wherein*, *whereto*, on or to the upper or furrounding fides---*elfewhere*, on the upper fide place or firma-
ment---*acrofs*, *afkant afkew*, *awry*, *afiant*, *aflope*, *athwart*, *crofs-
wife*, *traverfe*, *oblique*, the fpring water or river fide place which
is a flope or from an upright line-- *apart*, *afunder*, *afide*, *befides*, *feparately*, *feverally*, *apiece*, the lower part or divifion or
the fide fpring water part which fevers countries --*every where*,
the parts extended, exifting and fpringing---*no where*, their negatives.

Of time or motion---*when*, *whence*, the fpringing and its
fide---*how long*, *how often*, the fpring or action length or fide
round ;---*once only*, *only but*, one fide, quality or action---*twice*,
fecondly, the fide fpring of two fides action or quality---*thrice*,
thirdly, the fide fpring, its divifion and quality---*four times*,
fourthly, the furrounding fpring parts or fides and qualities---
often, *oftentimes*, *many times*, the circle or furrounding fides or
parts---*fo often*, the fide furrounding water part circle-- *yefterday*, the action, divifion or day paft-- *erft*, at firft in time for
fpring paft---*formerly*, *beforehand*, *beforetime*, *heretofore*, *aforetime*,
yore, *agone*, *long ago*, *a great while ago*, *laterally*, the lower fide border
or paft fpring--*betimes*, by the furrounding fide---*now*, *already*,
ever, *evermore*, *always*, in fpring, on the fpring fide, the fpring,
the furrounding fpring and on the way or fpring---*never*, the
negative of fpring---*to day*, *tomorrow*, *in the morning*, in or at
the divifion or furrounding fpring --*anight*, the no light fide or
divifion---*next*, the inclofing water fide---*ftill*, *yet*, *alfo*, *likewife*, *item*, *alike*, *again*, *eftfoon*, *encore*, *afrefh*, *anew*, *while*,
well nigh, *almoft*, an emanation of light from above on the
lower parts, fpringing up from the furface and on the furrounding fpring---*anon*, *forthwith*, *immediately*, *prefently*, *quietly*,
foon, in motion, a fpring from the fide me at the fide place,
before fent for or the upper light---*henceforth*, *henceforward*,
hereafter, on the fide in forwards --*then*, the fide, poffeffions, or
things in---*thence*, *thenceforth*, *thenceforward*, have been already
explained---*foon after*, after one furrounding fpring or fun—
fomewhile, *awhile*, *fometimes*, *long while*, the furrounding fide
fpring either of water or light—*feldom*, the fea water place border furrounding—*prematurely*, *overfoon*, before its fpringing to
the fide or furface—*annually*, *yearly*, the fpring upon alternately,
one after another—*at length*, *laftly*, *finally*, *ultimately*, at the fide
inclofing water place or other fpring fide or edge.

Of

Of quanties, qualities, &c.—*how many, how much?* the
upper parts or furrounding fprings—*fo many, fo much*, the furrounding parts or fprings—*more*, the fea or furroun ing water
—*lefs*, the fide fpring water place—*abfolutely, wholly*, the
furrounding fpring water place---*adjectively*, acting at the
fide water or other place with a fubftantive as a quality---*affec-
tionately, paffionately, pathetically*. at the fide part place by which
qualities, energies, and affections are moftly exprethible and ex-
preffed---*agrily*, action had or paft fide; *articulately*, the found
of the joints or articles-- *bodily*, in or like the body---*circum-
ftantially*, like the furroundiug things---*coldly*, like the fhut
water place, divifion, or fide --*fiercely*, like the action of fire --
naturally, like the internal fpring of things --*wholly*, like the
fpring---*worthily*, as the value of man or the fpring---*as*, the
earth's fide, furface or feen part---*why*, the fpring---*therefore*,
before the part upon--*adieu, hail, farewell*, at the fpring, flowing act, and the fpring place---*amen*, for in or in the furrounding parts or exiftence : and the other forts of adverbs are implied in the foregoing.

Adverbs are compared like adjectives by adding *er* to the
comparative, and *eft* to the fuperlative degrees, or letting *more*
and *moft* in oppofition thereto.

An example of their conftruction ; when adverbs *are locally*
or fixedly at the fide of verbs *they generally exprefs* qualities,
actions and attributes, but with adjectives *they feldom mean* any
thing more than the degrees of comparifon and quantities, as
he moveth faft, *indifferently faft*, or more or very faft, or he
fings fweetly or exceedingly fweet, but not fweetly fweet ; and
adverbs like verbs exprefs fubftantively, actively, and paffively,
but the properties of adverbs are beft difcovered at the fide of
verbs or adverbs according to the import of the name.

Of CONJUNCTIONS.

Conjunctions are active parts of fpeech which conjunctively
or disjunctively connect two or more names, propofitions or
fentences into one fentence. The former fort abfolutely connect them with their meaning, or only continue the fentences
without any compatibility of meaning, and the latter do fo
conditionally, exceptionably, or interrogatively

The firft forts of which are *and* or end, on in divifion ; *but*,
by the fide or things feen ; *yet*, it fprings or is ; *alfo*, on fo or
the furrounding tex ; *likewife*, the fame way ; *ftill*, the light
flowing on the lower fide or things ; *although*, upon the border ; *notwithftanding*, not oppofing the former action ; *however*,
the fpring or action as it may ; *neverthelefs*, without any fpring ;
as, the earth feen part or furface : *as well as*, the furface and
fpring

spring place below or under it ; *for*, the circle or border part ; *therefore* ; before the part upon ; *wherefore*, for the above cauſe or circle ; *becauſe*, by the cauſe or our own actions ; *that*, the ſide or thing at or towards. The disjunctives are *or*, the border or circle part, which may belong to the one ſide or the other ; *nor*, its negative or neither ſide ; either, the ſide ſpring water part, which may belong to either ſide ; *neither*, the negative of both ; *till* or *until*, during the ſide or ſurrounding ſpring of light : *while* and *whilſt*, the upper lights flowing on the lower parts : *if*, the flow of i or life ; *unleſs*, the ſpring out or leſſened ; *except*, the incloſing water ſide ſhut out of ; *beſides*, by the ſides

An example of their ſenſe and uſe in concord ; *if* our learned philoſophical, analogical, *and* moſtly approved Engliſh grammarians, had *as* I have diſcovered the primitive meaning of names, they *as well as* I might have been ſenſible *that* all the parts of ſpeech *or* names were ſignificant of things. *But although* their own good ſenſe and ingenuity might have inclined them to think favourable of my analytic inveſtigations, *yet* they, like the great Locke and others, ſuffered themſelves to be miſled by the ancients, who were themſelves *alſo* miſled *either* by the arbitrary and corrupt ſtate of their own languages *or* their prejudice againſt the revelations and opinions of ſuch as they beheld with contempt, *though* neither could juſtify the ſuppreſſion of their uſeful attention to the evidence produced to prove the divine origin of ſpeech, *notwithſtanding* their ſuperior knowledge or ſtations ; nor can any revelations be the leſs credible merely on account of its being delivered by poor ſimple and illiterate men, *unleſs* their ignorance, folly, prejudice or falſity were detectable; *neither* ſhould the doctrine of providence be otherwiſe deſpiſed or ſuſpected *except* in trifling inſtances, *as* Calphurnia's dreams and Brutus' viſions *while* they were aſleep, *becauſe* they might have been deceived by mere phantaſms. We ſhould *therefore* be cautious how we give credit to dreams, *likewiſe* to viſions, ſtill adhereing to ſuch well atteſted revelation as Moſes's account of the divine origin of ſpeech, which, if revealed, muſt give preciſion to ideas, and be the criterion of truth and knowledge. *Wherefore* theſe matters ſhould be well weighed before we expreſs our aſſent or diſſent thereto ; for we may not only ſuffer ourſelves to be deceived, but be the means of miſleading others contrary to the divine will thus manifeſted by his providence. And *however* trifling my former inveſtigations may be looked upon by ſome, it is certain *nevertheleſs* that they have been very differently received by many learned foreigners and ſome of our own country.

Of

Of PREPOSITIONS.

Prepofitions are fubftantive names which coalefce with, and are fet in appofition to fubftances and compofition with verbs as, *before a noun, prepofition*, to denote the cafes and relations as to place, time, and order of things, and to connect fubftantives in fentences, and difcourfes ; as, *from hence, with* fome reflection it may be inferred, that prepofitions are *by* themfelves fignificant *of* things ; and thofe of the Englifh language with their explications, are as follows :

In, man, or his poffeffions inclofed ; *with*, the fide fpring or a woman ; *within*, in the fide fpring ; *to*, the fide circle ; *into*, or *unto*, man in the fide circle , *at*, the fide ; *towards*, fpringing to the circle or border fide ; *of*, the circle or off part; *out*, the fpring circle fide ; *out of* the fpring circle off fide ; *without*, out of the fide fpring ; *from*, the furrounding parts or borders ; *after*, off the fide upon ; *behind*, by its fide ; *fince*, below the fide or action in ; *according t* , agreeing in the fame borders ; *between* or *betwixt*, the fpring part in the fide, or the inclofing fpring water part ; *among* or *amongft*, the things in or about the furrounding circle or borders ; *for*, the border part, which determined both the right and boundary ; *before*, by the border part ; *above*, the furrounding fpring part ; *up*, the fpring part; *on*, in the circle of place and motion ; *upon*, the fpring part therein ; *under*, the fpring in the earth of water, or the roots of plants not fprung up to the earth's furface ; *below*, the circle fpring water part ; *beneath*, by the water fide, down, on the border part *by*, the part ; *befides* ; by the fide ; *beyond*, by the inclofing circle or border fide ; *through*, the border water fide or from the fide ; *over*, from or out of the fide fpring water ; *over and above*, out of the river and furrounding fpring part ; *except*, the fide part fhut out of ; *until*, during the poffeffion of the fpring of light : and the two laft names alfo ferve for conjunctions,

Of INTERJECTIONS.

Interjections are certain energies or natural parts of fpeech or notes of exclamation of the feveral qualities and quantities or degrees of the human paffions, affections and emotions which are caft into or between the parts of difcourfes or fentences to denote the fame, and in moft languages they are fuch as follow :

Ah, eh, ih, o, ho, euge, which actively and paffively exprefs thofe of pleafure, rejoicing, laughing and exalting, according to the fenfe of the feveral words. And their tranfponents, ha, he he, hi, o, ho, hu, and heu, alack, alas, fie, woe, are expreſſive of pain, ſorrow, checking, calling, abhorring, and dejection of the paſſions. And they equally correſpond with the other parts of ſpeech in the meaning of their letters ; as, *woe*, from a ſpring ; *alas*, high low ; *alack*, from high ; *o*, from ; *uh*, an high ſpring ; *ah*, an high action ; *eh*, its reciprocal return ; *ih*, flowing high ; *oh*, the high circle or ſun, and the tranſponents are expreſſive of the contrary paſſions or their being ſhut out from thoſe or of their abſence.

Of SYNTAX.

The ſyntax of a particular language is the manner of rightly conſtructing words into ſentences, either ſimple or compound, according to the concord, government, and arrangement fixed upon and eſtabliſhed in that particular language. But that of a general and rational language is the form and mode of rightly arranging and conſtructing them according to their natural concord, government and meaning, or as they repreſent things and the order of nature.

The ſyntax of the primitive language was without doubt of the latter ſort; in which names of themſelves, without the aſſiſtance of art, naturally aſſumed their proper order and ſituation in diſcourſes and ſentences ; and in which ſtate they ſeem to have continued until their primitive meaning became gradually varied by the arbitrary rules of grammarians. But the Engliſh language having pretty well eſcaped the ill effects of ſuch unnatural refinements ſeems ſtill to retain ſo much of its primitive rational ſtate, as perhaps to ſerve for an univerſal language, without any other rules or diſtinctions of concord or conſtruction, than what has been here already indicated under the ſeveral heads.

And although we have grammars, like thoſe of the Greeks and Romans, abundantly ſtuffed with arbitrary rules and diſtinctions, the Engliſh language has not yet been bent by any arbitrary ſigns or rules of concord or conſtruction to the grammatical ſtate, ſo far as to ſerve for the inſtruction of youth, but it is evidently the natural voice and converſation of the teacher, and reading that inſtruct according to the genuine and natural principles, concord, and conſtruction of the language ; for we may diſcern in the works of our moſt approved writers, a great variety of phraſes and ſentences, which have been condemned or neglected by the grammarians, and yet as they coaleſce in concord, have been adopted by our very beſt writers. And it might perhaps be difficult to tell by whoſe grammar it was, that our celebrated Shakeſpeare learnt the Engliſh

tongue

tongue, and yet Shakefpeare's works contain many very expreffive and primitive names, concords, and phrafes incompre-
henfible to our modern critics and grammarians. In fhort as
the Englifh names ftill retain their primitive natural ftate, it is
the fenfe or nonfenfe, the truth or falfity and meaning of the
phrafe or propofition that directs its order or manner of conftruction.

It is however neceffary that the parts of any language agree
in concord with one another as to cafe, gender, number, and
perfon.

Now as to cafes, the Englifh language expreffes them by
prepofitious fet in appofition to fubftantives and in compofition
with verbs, and the bare pofition of fubftantives before and
after verbs, as to denote the nominative cafe or the fubject of
a propofition, its pofition ought to be before the verb, and after
it, to exprefs the objective cafes, except the verb neuter, as
has been already obferved.

As to the fexes or genders of fubftantive names, they are
diftinguifhable into the mafculine and feminine genders, only
as to animals and fuch other things as feem to have any analogy
therewith, and thofe that feem to partake of neither fex, are
named the neuter gender. And nouns adjective naturally correfpond with the fubftantives, as to genders as well as numbers without any other variation than to exprefs the degrees of
comparifon, as do the relatives, *who*, *which*, and *that*.

Nouns and pronouns naturally agree in concord with verbs
without any variations, and to lay down rules and diftinctions
without any real difference, tends to puzzle and confound rather than illuftrate languages. And as we have already under
the proper heads fully explained the connection and conftruction
of the feveral parts of fpeech in fentences, had not our grammarians entered upon a minuter detail thereof than was probably neceffary, we might perhaps here fafely take our leave of
the fubject, and leave our very primitive language to flow naturally in its antient courfe, but the prefent ftate of the Englifh grammar obliges us to add the following remarks on the
rules of conftructing an Englifh fentence.

With refpect to cafes, it is obferved, that a fubftantive before
a verb, except it be of the infinitive mode, muft be in the
nominative cafe, as *he writes*; and that fuch verbs muft always
have the nominative cafe either expreffed or implied; that the
fubftantive after the verb paffive or neuter, as, *you are*, *may
become*, or *be accounted a man*, muft be the nominative; but
after a verb active or tranfitive, as, *I delivered the key*, it muft
be in fome of the objective cafes; that the prepofition is often
underftood,

underftood, [as after give and before the nouns of time-that when a thing belongs to another, as, *Alexander's horfe*, or the *horfe of Alexander*, the poffeffor muft be in the genitive cafe, and to which fome other noun muft always belong—that when a fubftantive is added to explain another fubftantive, as, Alexander the king, they muft be both in the fame cafe by appofition-that the verb *to be*, has always the nominative cafe after it ; and that the nominative cafe following an auxiliary or other verb fupplies the place of the conjunctions *if* and *though*, as *had he done*, for if he had done fo. Now fome of thefe diftinctions of cafes may perhaps be proper as they are here explained ; but whatever foundation there may be for them in the Greek or Latin, or any other language whofe fyntax, as well as parts, are intirely grounded on arbitrary principles, the more rational Englifh, whofe only figns or diftinction of cafes, are exprefs prepofitions or thofe in compofition with verbs and the bare pofition of the fubftantives in concord with verbs, can receive but little advantage from fuch arbitrary or nominal diftinctions or modes of conftructing an Englifh fentence.

Then as to the genders and numbers of nouns fubftantive, it is remarked, that a noun of multitude may have a verb or pronoun agreeing with it either in the fingular or plural number according to the import of the word as expreffing unity or plurality, as, *my people is or are* foolifh, and the affembly of the wicked have inclofed me-that nouns in the fingular number joined together by conjunctions copulative have verbs, nouns and pronouns agreeing with them moftly in the plural number, as Socrates and Plato *were* wife, but if the fingulars fo joined together are of feveral perfons, in making the plural pronoun agree with them in perfon, the fecond perfon takes place of the third, and the firft of both, as *he* and *you* and *I* won it at the hazard of *our* lives; *you* and *he* fhared it between *you* ; and when the fubject is a thing without perfonality it is expreffed by the neuter pronoun *it*, as *it* happened or defcending. So that thefe rules tend to fhew that the Englifh genders and numbers are diftinguifhable only by the meaning of things, without any variation of terminations or other figns than what have been already intimated.

Adjectives and participles — an adjective after the verb paffive or neuter is faid to be in the nominative cafe—adjectives or participles are added to or put before fubftantives to exprefs their qualities, and in that cafe participles have the nature of adjectives—adjectives have fometimes the infinitive

* E

mood

mode after them, as, *worthy to die*—adjectives have no variation of genders or numbers, and some of them are derived from verbs. The distributive pronominal adjectives, *each, every, either,* agree with the nouns, pronouns and verbs of the singular number only, unless the plural nouns happen to be collectives-every adjective has relation to some substantive either expressed or implied. Adjectives sometimes become substantives and substantives adjectives, as do verbs become either and either verbs, so as to go along with the ideas of the various forts of substances, qualities, and actions. But it is farther observed as to participles, that with prepositions before them they answer the Latin gerunds, as, *by doing*; or, with the articles before and prepositions after them, they become substantives, as, *by the observing of* which we should erroneously suppose persons and participles to be gerunds and the concurrence of a preposition to be necessary to the article's substantiving power.

Verbs; a verb following another verb must be in the infinitive mode, as, *I hope to live*; *to,* before a verb is the sign of the infinitive mode. The verb in the infinitive mode often stands absolute or independent of the rest of the sentence, as *to conclude-two* negatives make an affirmative.

Adverbs; adverbs are added to verbs and adjectives, to express qualities and other circumstances, or the same may be done by a substantive and preposition; as, I am very lonesome or by myself; or, by comparing the same quality in different subjects, as, white as snow, wiser than I; adverbs have no government, but ought to stand at the side of verbs and adjectives according to the import of their names.

Conjunctions and relatives; two forts of words connect sentences, namely, conjunctions and relatives; it is said that conjunctions have the government of modes, but it seems more probable that they are adverbs of time; conjunctions have their correspondent conjunctions. as although-yet; whether-or: so-as; fo-that, neither-or; the relatives *who, which, that,* must have their antecedents either expressed or implied. and having no variation of genders or numbers must agree therein with their antecedents; *who* referring to persons, *that* to things or persons, and *which,* to actions as well as persons and things, and the relative is of the same person with the antecedent, and the verb agrees with it accordingly; and when this, that, these, those, refer to a preceding sentence, *this* and *these* relate to the later member or term, and *that* and *those* to the former.

Interjections; interjections have no government though usually attended with nouns in the nominative case and verbs in the indicative

indicative mode ; yet the cafe and mode are not influenced by them, but determined by the nature of the fentence ; or per- haps thofe forts of names may be more accurately defined to be adverbial names of the various degrees and circumftances of the paffions with which the fentence is conceived or expreffed or little abfolute or independent fentences of the paffions.

· The grammatical points ; as a period ought to be a compleat reprefentation of the water part circle or diftrict, it fhould confift of four fides or members, but whether thofe members or inclofing fides fhould be, , ; , . or , ; : . may for me be left to the printer's option.

R E M A R K S

ON THE

C I R C L E S OF G O M E R.

THE writer of the present essay once intended to engage in a more perfect delineation of the Circles of **Gomer**, to fix all the Celtic dialects, and thence investigate the religion, learning, laws, governments, tenures, customs, and manners of the Celtic or European nations. But finding that the specious structures of the antients, which mislead us in our modern superstructures, would require a more elaborate discussion than the reception of his past labours and present state of health will admit, the present remarks are designed more for engaging others in the pursuit of those important subjects, than a full and perfect illustration of them.

The principles on which the foregoing definitions are founded pre supposing the divine origin of human speech, it may not perhaps be improper here in the first place to observe, that most of the antient philosophers, particularly the metempsychosists, were of that opinion; as Moses must have been by recording that God himself formed man with a living, speaking, or communicating soul, after his own likeness, that he named the celestial and terrestrial created parts previous to the formation of Adam, and superintended Adam in giving names to beasts and birds. And as it appears by the foregoing definitions, that those names are analogically transferrable to all the other systems of generation, vegetation, and forms of things, according to the natural but inexplicable powers of letters and names in their affociation, correfpondence and coincidence with ideas in their various and moft extenfive fhapes, forms, and changes; languages, however in the compound they may partake of man's art, muft in the abftract be the gift of providence to Adam, and perhaps republifhed to Mofes on mount Sinai, or come to him by tradition; fince letters are defcriptive of his hiftory of the creation, and fall of man, and many other branches of original knowledge and philofophy, with which the Jews were at firft but little acquainted; and it is admitted, that philofophy took its rife amongft the Barbarians, Gentiles, or Gauls, probably from the Druids ufe of the inflecting form of confonants, by which they framed their fongs.

A language that is thus founded on revelation, and tending to the difcovery of original knowledge, muft, when properly and accurately defined, alfo tend to a precifion and reunion of languages, ideas, and opinions, and be the criterion of truth, and decifive in all cafes, fince the archetypes of language and original knowledge are the fame, and a right definition of the one muft explain the other. Whether thefe fheets contain fuch a difcovery is fubmitted to the candid examination and confideration of the learned. It feems however probable, that no other language can make fo near an approach to the firft univerfal language, and to natural precifion and correfpondence with ideas and things, as the Englifh in the form and mode it has been here introduced as an univerfal language; for in fhort the author has difcovered the firft language, and was it not for the fatigue, labour, expence, and fome other reafons of a more important nature, could as he imagines reftore it, and the foregoing definitions and grammar have been from thence formed.

If any other languages fhould, like this correfpond with the primary names of the elements, their divifions in place, and the hieroglyfic and argrafic prototypes, as the fea and river, the vulva and penis, and the mouth and tongue, or voice, fhould have naturally grown with ideas and things by virtue of the inflectory confonants, fhould be analogically transferrable to every fyftem of creation, and fhould excite and exprefs the very paffions, they may be brought forth in their beft form of univerfal grammar. And if on examination they fhould appear to be as perfect copies of nature as the Englifh language; then, as all fuch languages muft nearly correfpond, they may be confolidated into one univerfal language. But if no fuch can be produced, furely the beft form of the once univerfally fpoken language, ought to be generally received as fuch, as

having

having a tendency to the reſtoration of ancient knowledge and the reunion, intercourſe, tranquility and happineſs of mankind, and more perfectly ſerving the ends of providence in the affairs and government of the world. But thoſe that would be competitors to the Engliſh for originality or univerſality, ſhould firſt well conſider the imperfections of moſt dialects of the firſt language, and whether theirs, like the Engliſh, can bear the open light without being ſubjects of ridicule.

The pretenſions of the Engliſh language to univerſality, may be ſtill further urged from its being in the abſtract the mother of all the weſtern dialects and the Greek, elder ſiſter of the orientals, and in its concrete form, the living language of the Atlantics and aborigines of Italy, Gaul and Britain, which furniſhed the Romans with ſuch of their vocables as are not of a Greek origin and their grammatical names ; and from its impreſſion of the primary names of places on moſt parts of the globe. And though ſome of the Engliſh vocables may have been loſt and recovered by means of the Latin tongue, even thoſe, on a due examination, will appear to be radical Engliſh names borrowed by the Romans from the language ſpoken by the anceſtors of the Engliſh in Italy, Gaul, and Britain. And notwithſtanding the various attempts which have been made by different writers of aboliſhing it, and introducing other dialects in its ſtead, as the national language of thoſe countries in order to favour their uſurpations of the Celtic poſſeſſions, all that could be effected was to ſplit it into various arbitrary dialects, wilfully and corruptly deviating from the Engliſh, to ſerve as the living languages of ſome of thoſe countries.

Having in my former eſſays on the origin of knowledge and language, already ſhewn how the Celtic dialects and knowledge derived their origin from the circles of Triſmegiſtus, Hermes, Mercury or Gomer, we ſhall here only add a ſingle remark to ſhew how the Engliſh language happens more peculiarly to retain its derivation from that pureſt fountain of languages. In other languages then the nominal or literary characters as well as their ſounds, like thoſe of arithmetic, algebra, geometry, aſtronomy and muſic, ſeem to be arbitrarily applied, but in the Engliſh they ſignify and are applied as real characters, repreſentations, and expreſſions of nature by their ſymbolical figures and forms, and arithmetical proportion of ſounds or diviſions and multiplications of lines or muſical chords agreeable to the Circles of Gomer ; for Gomer ſuppoſed the world to be a large circle compoſed of many homogeneous leſſer circles, ſyſtems, or combination of elements, ſides, ſurfaces, figures, forms and parts. And ſuch as will deliberately examine the foregoing definitions and grammar of the Engliſh language, will find it to be an exact copy or defcription of the circles or fyftems of Gòmer. Hence alfo its greater pretenfions to univerfality.

The Englifh language alfo in virtue of the fpringing or inflecting powers, of its radical and mutable confonants, which muft have affifted the memory of the Druids in thofe various and copious fubjects which they taught, making a natural progreffion from its primitive ftate of nature to its prefent more complex or copious form, mode, and conftruction, and preferving the fame principles, and the like natural connection betwixt its feveral grammatical parts, as originally fubfifted betwixt particles and letters in the compofition of names, from the letter to the moft compound fentence, fo as to form an exact copy of nature, prevented its coalefcence with or admiffion of any foreign, exotic, or unnatural plants, and its being forced to the Latin or any other arbitrary language by any ufelefs rules of cafes, genders, declenfions of nouns, common and proper names, tenfes, modes, and conjugations of verbs, the government of nouns and verbs and many other diftinctions, with which it has little or nothing to do. Thus this antient language perpetuated itfelf, and preferved its originality and univerfality, and retains a peculiar property of explaining, not only other languages and names of perfons and places, but alfo the various branches of original and natural knowledge, fo as to make its way into the prefent ftate of the arts and fciences, and other matters of antiquity.

To produce every fpecimen of its powers in the invefligation of things, which may be given from the foregoing definitions, and to connect and apply thofe definitions to the origin of languages, nations, and things, would require at leaft 20 larger volumes than the prefent, we fhall therefore only inftance a few. If then we obferve the fcope and tendency of this infpired language, we fhall clearly fee that it was to be man's guide in the conduct and management of himfelf and the world, and that his chief concerns and duties here, were to be the propagation and prefervation of his fpecies, the cultivation and government of the world, and the exploration and adoration of the creator according to that one eternal and univerfal law or religion, which at fundry times has been promulged to man ; but that man by his fall from his original ftate, became of himfelf incapable of thofe purfuits. To fupply the defects and decays of human nature, letters and names, were to be to him and his pofterity as figns or fymbols for their conduct and better government of the world. Thefe feem to direct them to live in pairs in the ftate of propagation, to divide and govern the world in

communities

communities by meres, fences, and boundaries, which were to be the title and defence of separate poffeffions, and their neceffary fubdivifions and fubordinations, for the well being of communities, and to felectand fet apart a portion of time, place, and men, for the exploration and adoration of the caufe of all things.

Hence the origin of the connubial and governmental ties, and the divifions and fubdivifions of the earth into cantons, commots, paguffes, and leffer portions, amongft tribes and clans, under the government of patriarchs, druids, bifhops, earls, and other heads, and their fubftitutes. And to the grafting or coalefcence of many of thofe tribes under one head, and fubftituting the connection, ties, and privity of eftate, inftead of thofe of blood, which fubfifted amongft tribes under the patriarchal mode, every well conftituted monarchy owes its origin ; all elfe being founded upon violence, fraud, ufurpation, tyranny, and oppofition to the defignation of Providence.

Hence alfo the origin of the feodal tenure and government, the freeholder, artizan, and freeman's, as well as the prelate and barons, fhare in the legiflature, on the fubdivifions of poffeffions and other properties acquired, the periodical parliamentary meetings without fummons to form capitularies, the origin of bifhops, earls, headburrows, and tything-men, or the judges and juries, fworn to declare the truth, and the frank pledge, or liberty of paffing and repaffing the boundaries, which prevented the difagreeable neceffity of perfonal arrefts, and long imprifonments for debt, for want of fpeedy trials. And though thefe priviledges, as well as the ancient feodal government, have been altered or abolifhed in France by Charlemagne, who, having feized the ancient fiefs into his own hands, varied them to a fpecies of benefices, which the Normans introduced into England, they have been with general confent of prince and people happily preferved, regulated and confirmed in Britain by our wifer Alfred and his fucceffors, by the moft facred ties, according to the concomitancy of power and poffeffion, and the anciently approved Englifh maxim, that what concerned all ought to be treated by all, however the fervices annexed may have been here varied. And the like tranfitions from the original patriarchal mode of government to the feodal, or monarchical, have been made by the Jews in the days of Saul, according to 1 Sam. chap. viii. and afterwards in Greece, Italy, and ancient Celtica. Nor did the Romans abolifh the clan mode in Britain, but permitted the Britons to enjoy their ancient government, under the comites, earls, or knights of the landmen, and aldermen of the feamen, inftead of the druids, whom they expelled. And as it appears by the foregoing effay, that our tenures and

* F

government

government are founded upon, and coexiftent with the original divifion of countries, according to the fcheme of Providence in the government of the world, to deviate from this plan muft be direct treafon againft the Author and Governor of all things.

Though the clan or patriarchal mode of government generally prevailed in Britain till the Saxon heptarchy, it feems probable that there was fome deviation from it towards an ariftocracy, before the arrival of the Romans in the country; for the powerful clans of the Coritani, *the fpreading borders*, according to the Roman policy, of encouraging foreigners to become Roman citizens, in order to extend their empire, had extended their borders far beyond their original limits, into the countries of the Cornavii and Brigantes, and at laft founded the united ftates or cantons of the feven Burgenfes or Loegria, whofe metropolis was Leicefter, and afterwards the kingdom of Mercia, and finally that of Great Britain; and thence the ftory of Geoffery of Monmouth, of Locrin, Albanact, and Camber, the fons of Brutus, after their father's death, dividing the kingdom of Britain in three parts, called Loegria, Albania, and Cambria. It was from this union of the Britifh clans and their dialects that the Englifh language and nation chiefly derive their origin, as appears by innumerable proofs in the foregoing effay, and the following fpecimen of primary radicals, varioufly fpringing from the mutable confonants, and made ufe of in different parts of Britain as appellative names of places, according to their feveral qualities and fituation, and the bafis of different dialects, which affociated in the country and language of the Coritani, fo as to form the prefent Englifh dialect and local appellatives, and prove the prefent Englifh to be the genuine progeny of the original planters and land cultivators of this ifland. Nor are fuch radical names to be met with in any other country, neither do they differ from the Celtic meaning of things, or from one another, by any arbitrary, irregular, or foreign mixture or otherwife, but as fome are more compounded or tranfpofed than others, and thus the reft of our names feem to correfpond. But, though the people and language of Britain thus clearly appear to be of the fame Celtic origin, yet it is as evident that fome of every Celtic fpecies are in poffeffion of every country in ancient Celtica, whofe names of perfons and places, invariably made ufe of, and as appellatives, alluding to the radical or generical names, clearly diftinguifh them. And the following fpecimen alfo confirms the foregoing definitions of letters, as for inftance, le and pe, the water place, and water part of man and earth, or the penis flowing between the legs and a river.

ENGLISH.	DANMONII.	DUROTRIGÆ.
A part or water part; A p'ace or water place	Pe, peth, par, ba, ber; le, ley, pli	Pe, bere, bri, le, ly
A fea; a pool or lake	Mor; pool	Mor; pool
An haven, harbour, port or water gate	Aber, port, dor, mouth	Aber, port, dor, mouth
A well or fpring, a river or fide river; a vale or river fide	Bur, au, auc, ex, il, low; comb kil, glin, hamp	Au, auc, car, ftur; wey; coomb, ham
A way or road; a ford or water way; a bridge or ferry; a river bank or lane	Ford; rhyd; pont; lan or lang	Ford; rhyd; blan lan or lang
A border, mere, coaft, edge or confines	Kir, mar, mod, tor, min, pen, tot, egg	Cor, egg, fhir
A field, park or other inclofed piece of ground	Cae	Ca
A country, neighbourhood, or habitation	Bro, bod, land	Bro, fro, bury, burn, tir, bod
An houfe, hall, or palace; town, burrow; city	Tu, lus; tre, ton, fton; kaer	Don, ton; chefter
A ftone, rock, bank, hill, promontory or head, mountain	Carrock, carn, pen, hill	Crek, ftone
A grove, wood, foreft	Coet, cot, wot	Wot, Wood

BELGÆ.	ATTREBATÆ.	REGNI.
Ber, man, mon; le, bel, bley	Beth, ber, far; la, le, ley	Pet, beth, man, dar; le, leigh, ble
Mil, mer, fea, fhe, fo		Mar, fea
Aber, por, let, port, mouth	Let	Gate, mouth
Au, auc, ex, ver, les, lu, ly; coomb, chill, cham, ham, fom, dol Ford; rhyd, rith; lon, lan, lang	Fer, ver, wan, white, well, ock, maid; coom, ham, comp Ford; read; bret, long, ling	Bur, ver, guil, au, auc, oke, ote, wa, van; coom, den, ham, dale Ford, wick; lan, ling, clan, fus, futh
Cad, can, chard, crock, greke, lock, ing, ring	Beck, ing, lam, coln, ox, fhri, fhell	Chor, hor, cher, grin, mer, ing, fin
Cae	Cae, field	Cae, cha
Bro, bri, bury, ter, land, worth, bod	Bro, bri, bury, bourn, worth, land, fy	Bro, bri, bury, burn, ter, land, bod, worth
Tu, ftu; tre, don, fton; chi	Ton; chi	Ton; chi, chefter
Lech, ceric, pen, dun	Lech, dun	Ban, down, hill
Perth or berth, fel, wythie, wyde or wood	Wood	Cote, wood, hurft

CANTÆ.

CANTÆ.	DOBUNI.	CATTEUCHLANI.
Be, de, dar; le	Be, bar; le, ley, cley	Be, bea, ber, fer; le, ley, leigh, cley
Mer, mil, fhe	Sea, fe	
Dor, dover, let	Dor	Port, gate
Dur, au, auc, wye, went, well, maid, fever; coom, ham	Lyd, avon, whit, wes, puck; coom, den, ham	Win, oufe, euch, oak, buck, ware, lu; bal, den, ham
Ford, wick; rith, fhut; lang, ling, fand, fend	Ford, wick; rhyd, olaneag, ald	Ford, wick, ftrat; lan, lang, fulon
Can, ken, cow, ox, hor, mere, ing, folk, gren	Cher, kir, fcier, chad, chep, glo, ingle, ring, thorn, ftoke	Mere, ing, catti, dock, her, grave, green
Che, chae	Cae, cha, field	Cha, field
Bro, bury, born, worth	Bro, bri, bury, burn, lade, land, bi, bod	Bro, bury, burn, bi, bed, bod, worth
Ton, borrough; chi, chefter	Tu; ton; chi	Tu; ton; chaer
Stone, pen, hill, neffe	Lech, rich, ban	Dun, chil, ridge
Wood, hurft	Wood, hurft	Cot, wood

TRINOBANTES.

TRINOBANTES.	CORITANI.	CORNAVII.
Pri, bar, dra; le, lea, ley	Be, bra, der; la, le, ley, ble	Bar, bra, da, dar; man; le, ley, leigh, pel, bel, bli
Mor; lyn	Mer, fi, fey; lyn	Mer, fey; poll
Aber, porth, thor, thorp, gate	Thorp	Bret, port
Bur, dur, au, auc, brent, low, wall, hen, hun, ux; cam, comb, den, glen, dol, ham Ford; rith; long, lan, lang, fud	Bur, ver, brook, auvon, oke, afh, tow, wye, cley, fley, col, mow; den, dale, ham Ford; rud; pon, pont, bridge	Bur, dur, ftur, thurs, ons, wye, we, well, cley, cliff, han; comb den, dale, ham, val Ford, road, way; rhyd; bridge; lan, lang
Cro, gron, for, mer, ing, ching, thorn, ox	Cor, gra, nor, mar, kim, mel, ing, thorn, keft, ax, wake, pen	Cor, wor, nor, ken, keft, coft, feck, ing, thorn, tam, pen, fin
Ca, che, fhe, field	Cae, ca, cha, field	Ca, cha, che, feld, field
Bro, bury, burn, bery, tir, land, by, bigh, worth	Bro, brun, ter, land, by, bod, wrth, bos, vill	Bro, bri, burn, bourn, ar, ter, land, by, bewd, bos, worth
Tre, dray, don, ton, burrough; car	Tre, ftre, don, ton; car, char, har, ches, cefter, chefter	Tre, dra, try, don, ton; char, har, ches, cefter
Brun, dun, down	Rock, dun, mont	Lich, brin, dun
Cot, walt, wood, hurft	Cote, wood	Cott, cote, wood

SILURES and DIMETÆ.	ORDOVICES.	BRIGANTES.
Peth, tre, bre, rad; la, le, ley	Ba, pe, ber, der, red, man; la, le, ley	Ba, bar, ber dar, man; le, ley, lay, ple, pley, pel
Mor, mil, og; pwl, lyn	Mor, og; pwl, lyn, lac	Mor; poole
Aber, dor, port or, porth	Aber, porth, dor	Aber, thor, thorp
Auc, ock, buc, ufk, fwa, yft, wye, wyth, fynon, weli, cil, hen; cwm, glin, den, hamp Ford; rhyd; pont, bridge; lan, lon, lane	Fynon, avon, wy, wyg, wys, wen, cell, dur, clyd, rug; bala, cwm, glyn, den, glas Ford or forth; rhyd, ruth; pont; kil, lan, lon	Bur, dur, thur, ure, ans, wig, ote, ous, ri, cald, mus; cum, chum hum, ham, gles, dale Ford, forthe; pont, bridge; lan, lang, lon, kil, cli
Cwr, garth, to, can, mar, mon, ken, nid, neath, pen, fin	Cwr, garth, to, mon, can, kim, mencu, pen, min, fin	Cro, kir, crack, mar, mafh, mit, mede, thorn, cope, nid, pen, pon
Maes, cae, che, field	Cae, maes	Ca, cha, fhea, field
Bro, bri, fraw, tal, tir, glad, by, bod, wrth	Bro, bri, fraw, tir, tal, glad, by, bod, wrth	Bro, bor, born, bra, burn, are, land, by, bod, wort
Tu, hus, hall; tre, ton; din, caer, car, cefter	Tu, letu, betus, lus; tre; dinas, car, char, har	Houfe, leeds; caer, char, fhir, chefter, cafter
Careg, maen, craig, carn, cefn, dun, bryn, moneth	Lech, careg, maen, craig, carn, bryn, rhiw, galt, moneth	Stone, roch, fcar, dun, mond
Coed, fil, wood	Lwyn, coed, den, kelle	Calle, wood

OTTADINI.

OTTADINI.	Gadeni, &c. or the lowlands.	Calledonia, or Highlands.
Peth, ber, tin, thir; la, le	Bar, bri, dea, dee, thir, ster, men; la, le	Ber, perth, bra, bardin, far, er, ray, ardoch, mar, man, manoch; la, le, lo
Mor, og, se	Se; loch, lough	Mar; logh, lin
Port, mouth, gate	Borth, broth	Aber, broth, dor
Wal, well, otte, kes, hope	Au, lau, lid, wig, esk, col, rox, ren, drum, teifi; clyd, cwm, don, glas, dale, fal, ham	We, us, cul, leave, buqu, mur, dou, drum, han; bala, comb kil, glen, dale, &c.
Wick; rut; bridge; lan	Ford, forth, way; frith; kyle, lon, lane	Strath, ruth, reud, kyle, futh, len, lang, bruach
Cor, nor, kirk	Char, craw, merch, mers, mait, ing, ox, pen, phin	Cro, kir, for, tor, ing, kin, can, ang, mern, falk, neth
	Ca	Ca
Bury, land	Fraw, land, bod, bot	Bre, ar, tir, land, fife or five
Caer, chester	Tre, ton; burgh; caer	Din; don; burgh
Dun, stone	Stone, rick, carrick, crag, uchil, dun, knoc, ceantir, nesse	Crag, carn, dun, hol, born, bryn, alben, ceantir, nesse, mont
Wood	Coil	Coil, calle

Having

Having thus fixed the etymological science, and given proper hints towards settling the origin of nations and original knowledge on a firm basis, we shall here add a few more cursory observations on the whole of the foregoing work. And, in the first place, we submit it to the consideration of the learned, whether since the etymological science and primitive meaning of languages and names of places and things, which are equally appellative, have been explained with certainty and precision, they will not clearly explain and distinguish the origin of all languages, nations, people and things, according to that analogy which appears to subsist betwixt all systems, and point out in particular the first planters of Europe, at least on this side the Vistula, Taurica Cherfonesus, Hellefpont and Gades, to be the Celtes or Gomerites, from whose language all the names of places, countries, persons, nations and dialects seem to be derived: and though some have imagined the Sclavonic and Teutonic to be European originals, if it may not be more probable that they sprung from a mixture of the Celtic and the ogum or fluctuating dialects of seamen, who had no concern in the cultivation of countries or original impofition of the names of places, since all the radical names of places, persons and dialects now appear to be of one and the same Celtic original. If then the English should appear to be the best Celtic dialect and the original language of the Europeans, why should it not be restored as the general language of those countries?

It is also conceived that the originality of nations stands in the same order as the Roman alphabet, which was probably fixed with that view at the extent of the Roman empire, and that their origin, confistent with their radical and derivative names, which are derived from the respective countries, must, like the names of countries and things, all correspond with the form of inflecting the radical and derivative confonants in the foregoing grammar, whereby m and ph or f appear to be derivatives of the radicals p and b, g, n and ch of c, and d, n and th of t; as for instance the Phrygians, Phryfians, Franks, Mauritanians, Merini and Marcomans must have sprung from the Beri, Beriges, Cum-beri and Britons; the Galli or Gauls, and Gomeri from the Calle, Kelle or Celtæ and Cwmbri, and the Thracians from Tyras the younger fon, offspring or nation of Japhet, and so as to those nations who have the auxilliary confonants for their radicals according to the explanation thereof in the said grammar.

Hence, confistent with venerable Bede, we may fafely conclude that the proper Britons, who were the first poffeffors of that part of the ifland called Britain or the fouthern part, and

the

the Brigantes of the northern parts migrated here from the Merini of Armorican Gaul, who came from the Iberi of Spain, the descendants of the Beriges and by whom a considerable part of Ireland was planted, and from whom the Scots and Saxons are mostly descended, but they were mostly seafaring people. That those were succeeded or rather accompanied by the land-men of Gallia Comata called Calle, Kelie or Celtæ, who possessed the river sides, vales and midland countries, leaving the sea coast to the Britons, Scots and Saxons, and from whom came the Caledonians and Picts. And that those were succeeded by the Cumbe i or Cumberi-Gauls of Gallia Bracata and Togata or the upper countries or borders, who had been incorporated by the first Hercules with those of the Comata, and from whom the Gauls, Welsh or Cumbri of the Iter Cymbrorum, Wales, and the countries betwixt the Humber and Tweede, and on the western side to Dunbarton, and other considerable parts of England, Scotland and Ireland, seem to be descended. And from a mixture of all these the present inhabitants of Britain and Ireland derive their origin. But there were the like migrations of those people over the Rhine into Germany and Denmark, so that those countries seem to be copies of ancient Gallia, as Gaul was of Thrace, Dalmatia, Italy and Spain, and those of the lesser Asia or Natolia and Greece; the first being more particularly the Titans, Frisians and Britons; the second, the Celtiberi and Calabrians, and the third, the present English, as well as most branches of the Danes, Swedes, Germans and French; but then it is to be understood that the latter comprehend all the former emigrants.

The Cumbri, of the Comata, and other Gauls, probably by the Druids, governed Europe, and divided the countries by rivers and motes into commots and cantons, according to the form and disposition of Eden and the capital Roman letters, and always regularly dwelt in cities; leaving the open spaces betwixt the borders of the respective cantons and the sea coasts, as forests and heaths for the use of the nomades or shepherds, seamen and merchants. And their migrations were made and vicinity preserved regularly from East to West in streets from Asia to the remotest parts of Ireland, as appears from many circumstances and proofs in the foregoing and former Essays.

The incorporation of Hercules, which produced the Gauls or Cumberi, and of the Romans, who raised levies in the south to guard the northern parts, and sent others from thence to guard the south; and other subsequent intermixtures have caused an agreeable and pleasing variety, which still appears

in

in the very shapes, countenances, complexions and dispositions of the present English: But some of the Scots and Irish clans having for many ages avoided any intercourse or commerce with strangers, or even with each other, have contracted somewhat of a sameness peculiar to themselves, though a high-land laddy with a lowland lassy have produced many a bonny English bearn. And as the disputes betwixt the Scots, Picts, Saxons and Britons, on the retreat of the Romans from Britain, seem to have been no other than a civil war betwixt the same people and the Scots and English, appear to be of the same origin, and a greater intercourse and commerce betwixt the individuals of each country, might tend to a dissipation of the spirit of parties and prejudice and the national union, it seems more advisable to promote such an intercourse by every proper method than distress each other. And were leases of the wastes and uncultivated grounds in England and Scotland granted to such of the English and Scots as would intermarry, and sub-mit to reside in either country, and frequent church and kirk indiscriminately, all marks of distinction, as to arbitray rites and customs removed or conformed, and all opprobrious party names dropped, or somewhat more expressive made use of, as, Dons, Cons, and Ferrets, it might perhaps in some measure be productive of that end, and habituate the Scots to the ancient customs and common law of England, which differ somewhat from those of Scotland, as the latter par-takes more of the Roman than the former, and then we may perhaps be so happy as to differ only upon just and equitable principles. And though it was the will of Providence, in the colonizing age, to cause a dispersion of mankind by the con-fusion of language, the tendency of our Saviour's advice for a reunion, the growth of monarchies, the illustration and re-storation of languages, and various other circumstances in this age of cultivation and improvement, indicate the will of Provi-dence to be towards a reunion. And were mankind once brought back again to speak one language with precision, it seems probable that they would of course become of one opi-nion, and that Britons of all denominations would reunite under the most general name of English, or the island men.

And to prove the confusion of knowledge to be the effect of that of language, we need only recur to our late altercations with respect to the America government, the right of the collec-tive body to a representation in parliament and the privileges of the deputed members in their separate state over the collec-tive body, for want of a due conception of the first principles

*G 2

or

or original divine frame of government on which the English
conftitution was founded, but with which we might have been
fufficiently furnifhed by a right definition of the terms of our
laws and tenures and names of places and things: For, though
both parties muft have meant to fupport the public liberties, as
well as their refpective claims, each feemed to purfue a courfe
directly oppofite to the propofed end, namely, our colonies
ftruggled hard to fhake or get rid of the Englifh conftitution,
and to fubject themfelves to a charter or royal conftitution of
government, and the houfe of reprefentatives were as refolute
to prevent it, and at the fame time of fupporting an authority
over their conftituents. But had the origin and principles
of our government been inveftigated by the names of things,
which clearly fhew the natural and political divifions of countries and poffeffions and the original foundation of government
and order of the world revealed to our forefathers, as in the foregoing work, without doubt each fide would have purfued a
very different conduct for the prefervation of their refpective
rights and powers; fince they could not have imagined that
any party or body of Englifh people could make a conqueft
of the reft, or annihilate the Englifh conftitution, without the
mutual ruin and deftruction of the whole people and
country.

It would be endlefs to produce fpecimens of the powers of
languages in the inveftigation of things. It is fhortly the
author's plan in his feveral effays to inveftigate the origin of languages, nations, knowledge and opinions, fo as to introduce
an univerfal language on the firft natural principles of fpeech,
to reftore ancient knowledge, and reunite mankind. To that
end he has at leaft attempted to fhew that human fpeech derives
its origin from revelation, that letters, as real characters, as well
as reprefentations of founds, have a natural correfpondence
with, and, like names, are completely expreffive of ideas and
things, as their natural archetypes, of which larger names and
fentences are combinations, and as fuch the reprefentations of
the various fyftems or circles of things. Upon thefe princi les
all forts of names both of places and things, which are equally
appellative have been defined, and an univerfal Englifh grammar
framed agreeable to the original and natural conftruction of
fpeech, and the rational and proper arrangements of tenfe and
founds, in which the Englifh and other grammars, are of courfe
corrected. And after producing various fpecimens of the
powers of letters in the inveftigation of things in fupport of his
definitions, the origin, learning, cuftoms, manners and o. idicis
of people are in fome meafure explained by the names of

perfons, places and things ; and thofe fupported by the opinions of ancient writers.

And if we compare the names of perfons with thofe of places from whence the former are derived, and duly revife and rectify the annals of our country by the foregoing effay, we fhall be fu'ly convinced that the individuals have been fo well intermixed, that fome of every fort may be met with in every Celtic country ; and as fuch a connection mutually redounds to our advantage and glory, and tends to the reftoration of peace and unanimity amongft us, we fhould inftead of counteracting the ends of Providence as well as the author's endeavours, and wilfully and perverfely depreciating our feveral origin and antiquities, mutually concur in the profecution of the author's plan. But fo far are we at this time from fuch a purfuit, that not only the hacks, quacks and puffs of all countries, but alfo gentlemen, and fome of Englifh extraction, perhaps without knowing it, and merely to gratify vanity and national prejudice, or for amufement, bufily employed in feparating and depreciating the origin and antiquities of Great-Britain, and promoting divifions amongft mankind. Had our Fars in parfon, and its inflection with mac prefixed, fignifying the fon of parfon, the fon of Par, of Englifh origin, a little more deliberately examined the matter, our learned antiquaries, without doubt, would have feen that no diftinction in favour of the Scots and Irifh, exclufive of the Englifh, can be juftly deduced, either from hiftory or etymology, derived fromthe language of the firft poffeffors, or of thofe who impofed the names, but that whatever had a tendency towards leffening the one, muft depreciate the other, but thofe gentlemen have unfortunately reforted to the wrong dialect for their definitions of a few detached names, and looked upon our antiquities merely through the falfe medium of the ancients, inftead of availing themfelves as they ought, of our late publications. It is however to be hoped that our little effay on the Orphean ftrings, has at laft forced out the truth from things inanimate and cleanfed the Augean ftable, fo that we may all clearly fee its bottom, and convince all vain deluders of themfelves and others, that the origin and antiquities of the Englifh nation and language are inferior to none, and induce them to national love and fellow feeling, and to glory in the name of an Englifhman, and their near alliance to the firft people of Europe, the replenifhers, cultivators, and governors of the earth, and chief promoters of the wife ends of Providence , but as to thofe that may ftill feek after a fuperior origin and antiquity to the Eng ifh, they will at laft find themfelves to be in the purfuit of " a vain imaginary notion that draws

in

in raw and unexperienced men to real mischief, while they hunt a shadow *."

To conclude, we upon the whole conceive the origin, language, and antiquities of the celtic nations, particularly the English, to have been much misrepresented and injured by their own as well as foreign historians, antiquaries, lexicographers and grammarians. For, besides the various proofs already given of it, the following remark shews, that the people and language are aboriginal, viz. the people appear to be so by their language, and the language by its nearer approach to the original language than any in Europe, or perhaps elsewhere, and its having been framed by the hieroglyphic and argraphic archetypes, and the inflecting form of consonants by the druids or others concerned in the first division of the European countries agreeable to the original form of letters, and though the inflecting form of speaking has been disused, in order to introduce more complex names therein, its original use still appears by the regular progression of the radical consonants, in their variations from the names of substances to form those of qualities according to the inflecting form, and its being still preserved in speaking in its sister dialect the Welsh, and in the present state of the English vocables and connectives, so as to form a natural connection of grammar. And as the Roman characters only are intirely hieroglyfic and argrafic, it seems more probable that the Egyptians, Phoenicians, and the Greeks framed their adulterated and mixed characters by those of the Tuscans and Umbri of Cisalpine Gaul, the ancestors of the English, than that they borrowed their more perfect letters from those which want the hieroglyfic or secret part of the alphabet. But as to the use that was made of them, besides what has been already observed, and the preservation of their laws, mysteries and

---

* Orpheus signifies *the lower border water part or its guardian,* as does Euridice, *the lower spring water side* or *female* ; and Orpheu's journey to the realms of Pluto to set his wife Euridice free, means his going with Colonies to the western spring water parts to drain the country. He was master of Hercules, whose employment was of the same nature in cleansing Augeas's stable or *the draining spring water places,* by drawing rivers through them, or making drains for their discharge into the sea. They appear from various instances explained in the foregoing essay to have been two of the Phrygian Curetes, or Ida Dactyls, the distributors, protectors, and guardians of borders or countries and successors of Gomer, Pluto, and Mercury, to whom the Greeks and Romans gave the name of Dioscuri, the divine guardians. Of the same sort were the Argonautic expeditions.

religion

religion and the composition of songs, it does not at this time
clearly appear, though a very small volume written in those
characters might comprehend our huge bodies of English
laws, and be easily retained in memory by means of the connection of the inflecting form of consonants, preserved by the
poets in their compositions. And so far were the Coritani,
whose dialect it is from being expelled their country, that they en-
couraged the Cumbri, Romans, Germans, Scots, Saxons, Danes,
and others, who were the Norman ancestors, after the manner of the Persian, Macedonian, Roman, German, and
all other great empires, to make settlements amongst them,
in order to extend and defend their borders; and by
whose support they at first founded the Loegrian domi-
nions, preserved them in the Mercian and East Angles, and
compleated them in the English nation, and with whom the
Welsh or Cumbri of the Iter Cymbrorum, or Watling street, or
the Salloniacæ, who differed but little from them in point of
origin, whatever the old Britains might be, became consolidated
as appears by the marriage of Locrine and Guendolena. And
the distinctions of nations, as well as our disputes mostly arose
betwixt the sea race, as first possessors, and the landmen, as the
cultivators of countries, and amongst such as were allied to the
one or the other of them, who having no fixed settlements, made
depredations in those of others, and from downright usurpation
and tyranny, like those of the Medes and Persians, Greeks and
Romans, who were mostly the descendants of Madai and Javan,
the brothers of Gomer, on their elder brother's sons possessions.

And as our essay towards an union of languages and people,
is founded on the right principles of investigating the subjects,
and contains a just representation of things, it may at least
serve as a test of the real designs and purposes of my fellow
labourers, and the sincerity of our intention towards a coalition
and union; for those that are sincere therein, will approve of
every well meant step towards it, but others who entertain different views, will at least privately condemn it.

FINIS.

www.ingramcontent.com/pod-product-compliance
Lightning Source LLC
Chambersburg PA
CBHW020851270326
41928CB00006B/645